The Human Brain during the First Trimester 6.3- to 10.5-mm Crown-Rump Lengths

This second of 15 short atlases reimagines the classic 5-volume *Atlas of Human Central Nervous System Development*. This volume presents serial sections from specimens between 6.3 mm and 10.5 mm with detailed annotations, together with 3D reconstructions. An introduction summarizes human CNS development by using high resolution photos of meth-acrylate-embedded rat embryos at a similar stage of development as the human specimens in this volume. The accompanying Glossary gives definitions for all the terms used in this volume and all the others in the *Atlas*.

Key Features

- Classic anatomical atlas
- Detailed labeling of structures in the developing brain offers updated terminology and the identification of unique developmental features, such as, germinal matrices of specific neuronal populations and migratory streams of young neurons
- Appeals to neuroanatomists, developmental biologists, and clinical practitioners
- A valuable reference work on brain development that will be relevant for decades

ATLAS OF
HUMAN CENTRAL NERVOUS SYSTEM DEVELOPMENT
Series

The Human Brain during the First Trimester 6.3- to 10.5-mm Crown-Rump Lengths

Atlas of Human Central Nervous System Development, Volume 2

Shirley A. Bayer and Joseph Altman

CRC Press
Taylor & Francis Group
Boca Raton London New York

CRC Press is an imprint of the
Taylor & Francis Group, an **informa** business

First edition published 2023
by CRC Press
6000 Broken Sound Parkway NW, Suite 300, Boca Raton, FL 33487-2742

and by CRC Press
4 Park Square, Milton Park, Abingdon, Oxon, OX14 4RN

CRC Press is an imprint of Taylor & Francis Group, LLC

LCCN no. 2022008216

ISBN: 978-1-032-18328-2 (hbk)
ISBN: 978-1-032-18327-5 (pbk)
ISBN: 978-1-003-27061-4 (ebk)

DOI: 10.1201/9781003270614

Typeset in Times Roman by KnowledgeWorks Global Ltd.

Access the support material at: https://routledge.com/9781032183282

CONTENTS

ACKNOWLEDGMENTS

We thank the late Dr. William DeMyer, pediatric neurologist at Indiana University Medical Center, for access to his personal library on human CNS development. We also thank the staff of the National Museum of Health and Medicine, who were at the Armed Forces Institute of Pathology, Walter Reed Hospital, Washington, D.C. when we collected data in 1995 and 1996: Dr. Adrianne Noe, Director; Archibald J. Fobbs, Curator of the Yakovlev Collection; Elizabeth C. Lockett; and William Discher. We are most grateful to the late Dr. James M. Petras at the Walter Reed Institute of Research, who made his darkroom facilities available so that we could develop all the photomicrographs on location rather than in our laboratory in Indiana. Finally, we thank Chuck Crumly, Neha Bhatt, Kara Roberts, Michele Dimont, and Rebecca Condit for expert help during production of the manuscript.

AUTHORS

Shirley A. Bayer received her PhD from Purdue University in 1974 and spent most of her scientific career working with Joseph Altman. She was a professor of biology at Indiana-Purdue University in Indianapolis for several years, where she taught courses in human anatomy and developmental neurobiology while continuing to do research in brain development. Her lengthy publication record of dozens of peer-reviewed scientific journal articles extends back to the mid 1970s. She has co-authored several books and many articles with her late spouse, Joseph Altman. It was her research (published in *Science* in 1982) that proved that new neurons are added to granule cells in the dentate gyrus during adult life, a unique neuronal population that grows. That paper stimulated interest in the dormant field of adult neurogenesis.

Joseph Altman, now deceased, was born in Hungary and migrated with his family via Germany and Australia to the United States. In New York, he became a graduate student in psychology in the laboratory of Hans-Lukas Teuber, earning a PhD in 1959 from New York University. He was a postdoctoral fellow at Columbia University, and later joined the faculty at the Massachusetts Institute of Technology. In 1968, he accepted a position as a professor of biology at Purdue University. During his career, he collaborated closely with Shirley A. Bayer. From the early 1960s to 2016, he published many articles in peer-reviewed journals, books, monographs, and online free books that emphasized developmental processes in brain anatomy and function. His most important discovery was adult neurogenesis, the creation of new neurons in the adult brain. This discovery was made in the early 1960s while he was based at MIT and was largely ignored in favor of the prevailing dogma that neurogenesis is limited to prenatal development. After Dr. Bayer's paper proved that new neurons are adding to granule cells in the hippocampus, his monumental discovery became more accepted. During the 1990s, new researchers "rediscovered" and confirmed his original finding. Adult neurogenesis has recently been proven to occur in the dentate gyrus, olfactory bulb, and striatum through the measurement of Carbon-14—the levels of which changed during nuclear bomb testing throughout the 20th century—in postmortem human brains. Today, many laboratories around the world are continuing to study the importance of adult neurogenesis in brain function. In 2011, Altman was awarded the Prince of Asturias Award, an annual prize given in Spain by the Prince of Asturias Foundation to individuals, entities, or organizations from around the world who make notable achievements in the sciences, humanities, and public affairs. In 2012 he received the International Prize for Biology, an annual award from the Japan Society for the Promotion of Science (JSPS) for "outstanding contribution to the advancement of research in fundamental biology." This Prize is one of the most prestigious honors a scientist can receive. Dr. Altman died in 2016, and Dr. Bayer continues the work they started over 50 years ago. In his honor, she has set up the Altman Prize, awarded each year to an outstanding young researcher in developmental neuroscience by JSPS.

INTRODUCTION

ORGANIZATION OF THE ATLAS

This is the second book in the *Atlas of Human Central Nervous System Development* series, 2nd edition. It deals with human brain development during the early first trimester. The five specimens in this book have crown-rump (CR) lengths from 6.3- to 10.5-mm with estimated gestation weeks (GW) from 5.0- to 6.5. To link crown-rump lengths to gestation weeks (GW) we relied on ultrasound data shown in Loughna et al. (2009). These specimens were analyzed in Volume 5 of the 1st edition (Bayer and Altman, 2008). The annotations emphasize the continuing stockbuilding neuroepithelium (NEP) along the expanding shorelines of the brain's superventricles, the early shifts to a neurogenetic NEP (especially in the hindbrain), the continual growth of the superarachnoid reticulum, and the interactions between the brain and peripheral structures in the head and pharyngeal arches, especially as they relate to the rhombomeres and sensory cranial nerves.

The present volume features five normal specimens. Three are cut in the frontal/horizontal plane, two in the sagittal plane. Each specimen is presented as a series of grayscale photographs of its Nissl-stained nervous system sections including the surrounding body (**Parts II** through **VI**). The photographs are shown from anterior to posterior (frontal/horizontal specimens) and medial to lateral (sagittal specimens). The dorsal part of each frontal/horizontal photo is toward the top of the page, the ventral part at the bottom, and the midline is in the vertical center. All frontal/horizontal specimens have computer-aided 3-dimensional reconstructions of their brains showing each section's location. That reconstruction clears up the ambiguity about the exact plane of sectioning through each specimen. For all sagittal specimens, the left side of each photo is facing anterior, right side posterior, top side dorsal, and bottom side ventral.

SPECIMENS AND COLLECTIONS

The specimens in this book are from the *Minot and Carnegie Collections* in the National Museum of Health and Medicine that used to be housed at the Armed Forces Institute of Pathology (AFIP) in Walter Reed Hospital in Washington, D.C. Since the AFIP closed, the National Museum was moved to Silver Springs, MD; these Collections are still available for research.

Three of the specimens are from the Carnegie Collection (designated by a C prefix), that started in the Department of Embryology of the Carnegie Institution of Washington. It was led by Franklin P. Mall (1862–1917), George L. Streeter (1873-1948), and George W. Corner (1889–1981). These specimens were collected during a 40- to 50-year time span and were histologically prepared with a variety of fixatives, embedding media, cutting planes, and histological stains. Early analyses of specimens were published in the early 1900s in Contributions to Embryology, The Carnegie Institute of Washington (now archived in the Smithsonian Libraries). O'Rahilly and Müller (1987, 1994) have given overviews of first trimester specimens in this collection.

Two specimens are from the *Minot Collection* (desig-nated by an **M** prefix), which is the work of Dr. Charles Sedgwick Minot (1852–1914), an embryologist at Harvard University. Throughout his career, Minot collected about 1900 embryos from a variety of species. The 100 human embryos were probably acquired between 1900 and 1910. From our examination of these specimens and their sim-ilar appearance, we assume that they are preserved in the same way although we could not find any records describ-ing fixation procedures. The slides contain information on section numbers, section thickness (6 μ, to 10 μ), and stain (aluminum cochineal).

PLATE PREPARATION

All sections of a given specimen were photographed at the same magnification. M2300 was photographed with an Olympus microscope, C8966, C6516, and M1000 with a Wild Makroscop, and C8314 with a stereozoom microscope that was available at the Armed Forces Institute of Pathology. Sections throughout the entire specimen were photographed in serial order with Kodak technical pan black-and-white negative film (#TP442). The film was developed for 6 to 7 minutes in dilution F of Kodak HC-110 developer, stop bath for 30 seconds, Kodak fixer for 5 minutes, Kodak hypo-clearing agent for 1 minute, running water rinse for 10 minutes, and a brief rinse in Kodak pho-to-flo before drying.

The negatives were scanned at 2700 dots per inch (dpi) with a Nikon Coolscan-1000 35 mm negative film scanner attached to a Macintosh PowerMac G3 computer which had a plug-in driver built into Adobe Photoshop. The negatives were scanned as color positives because that brought out more subtle shades of gray. The original scans were converted to 300 dpi using the non-resampling method for image size. The powerful features of Adobe Photoshop were used to enhance contrast, correct uneven staining, and slightly darken or lighten areas of uneven exposure.

The photos chosen for annotation in **Parts II** through **VI** are presented as companion plates, designated as **A** and **B** on facing pages. **Part A** on the left page shows the full contrast photograph with labels of peripheral neural structures; **Part B** on the right page shows low-contrast copies of the same photograph with superimposed outlines of the labeled brain parts. The *low-magnification plates* s how entire sections to identify the large structures and subdivisions of the brain. The *high-magnification plates* feature enlarged views of the brain to show tissue organization. This type of presentation allows a user to see the entire section as it would appear in a microscope and then consult the detailed markup in the low-contrast copy on the facing page leaving little doubt about what is being identified. The labels themselves are not abbreviated, so the user is not constantly having to consult a list. Different fonts are used to label different classes of structures: the ventricular system is labeled in **CAPITALS**, the neuroepithelium and other germinal zones in **Helvetica bold**, transient structures in ***Times bold italic***, and permanent structures in Times Roman or **Times bold**. Adobe Illustrator was used to superimpose labels and to outline structural details on the low contrast images. Plates were placed into a book layout using Adobe InDesign. Finally, high-resolution portable document files (pdf) were uploaded to CRC Press/ Taylor & Francis websites.

3-DIMENSIONAL COMPUTER RECONSTRUCTIONS

This process took five steps. *First,* image files in the series for each specimen were placed into a Photoshop stack with each image in a separate layer. *Second,* by altering the visibility and transparency of these layers the sections were aligned to each other. After alignment, each layer was exported as a separate file. *Third,* Adobe Illustrator was used to outline the brain surface of each aligned section, and these contours were saved in separate Adobe Illustrator encapsulated postscript (eps) files. *Fourth,* the eps files were imported into 3D space (x, y, and z coordinates) using Cinema 4DXL (C4D, Maxon Computer, Inc.). For each section, points on the contours have unique x-y coordinates and the same z coordinate. By calculating the distance between sections, the entire array of contours was stretched out in the z axis. The C4D loft tool builds a spline mesh of polygons starting with the x-y points on the contour with the most anterior z coordinate and ending with the x-y points on the contour with the most posterior z coordinate. The spline meshes of the entire brain surface were rendered completely opaque at various camera angles using the C4D ray-tracing engine (**Figures 10** to **14** in **parts II** through **VI**). *Fifth,* in all frontal/horizontal-sectioned specimens, models of the brain surface posterior to a specific section were rendered with a copy of the photograph of that section texture mapped as a front cap on the model (*insets in* **Part A**).

NEUROGENESIS IN SPECIMENS
(CR 6.3-10.5 mm)

The specimens in this volume are equivalent to rat embryos on embryonic days (E) 12 to E13 based on our timetables of neurogenesis using ^{3}H-thymidine dating methods (Bayer and Altman, 1995, 2012-present; Bayer et al., 1993,1995). **Table 1** lists populations being generated in the spinal cord and medulla (**Table 1A**), the pons and cerebellum (**Table 1B**), the mesencephalon (**Table 1C**), the diencephalon (**Table 1D**), and the telencephalon (**Table 1E**). For all tables, the left panel lists neurogenesis in E12 rats (comparable to 6.3- to 8.0-mm specimens), the right panel in E13 rats (comparable to 10- to 10.5-mm specimens). Many of these populations are not distinguishable in the brain and spinal parenchyma, and often newly generated neurons are sequestered in the NEP before they migrate out (Bayer and Altman, 2012-present). We use methacrylate-embedded rat embryos at E12 and E13 to show the fine detail of the spinal cord (**Fig. 1**, Bayer, 2013-present), medulla, pons, cerebellum, mesencephalon, diencephalon, and telencephalon (**Figs. 2-9**). It is presumed that human specimens, if they were preserved in a similar embedding medium, would show the same features. Each table and figure set will be discussed briefly to summarize development in the spinal cord and different regions of the brain.

Table 1A: Neurogenesis by Region

REGION and NEURAL POPULATION	CROWN RUMP LENGTH	
	6.3-8.0 mm	10.0-10.5 mm
SPINAL CORD		
Cervical somatic motor	●●	●
Thoracic somatic motor	●	●●
Thoracic visceral motor	●	●●
Lumbosacral somatic motor	●	●●
Contralateral sensory relay	●●	●●
MEDULLA		
Retrofacial nucleus	●	●
Abducens (VI) nucleus	●●	●
Dorsal Vagal	●●	●
Ambiguus		●
Trigeminal (V, spinal continuation)	●	●
Hypoglossal	●●	
Gracilis	●	●●
Cuneatus	●	●●
Solitary	●	●●
Superior vestibular	●	●●
Lateral vestibular	●●	●
Inferior vestibular	●	●●
Medial vestibular	●	●●
Hypoglossal prepositus	●	●
Dorsal cochlear		●
Anteroventral cochlear		●
Posteroventral cochlear		●
Reticular formation rostral	●	●●
Reticular formation caudodorsal	●●	●
Reticular formation caudovenral	●	●●
Raphe complex	●●	●●

Table 1. Neural populations in the spinal cord and medulla (**A**) that are being generated in rats on Embryonic day (E) 12 (comparable to humans at CR 6.3-8.0 mm) and on E13 (comparable to humans at CR 10-10.5 mm). *Green dots* indicate the amount of neurogenesis occurring: one dot=a small amount (<15%); two dots=a larger amount (>15-90%). Populations with 2 dots in both columns means that substantial neurogenesis occurs in both time periods. This same dot notation is used for all of the remaining parts (**B-F**) of **Table 1** on the following pages.

TABLE 1A/FIGURES 1-2

All of these data are based on rats that have a similar morphological appearance to human embryos. It is assumed that similar developmental events are happening in the two species (Bayer et al., 1993, 1995; Bayer and Altman, 1995). Thus, the data in the left panel of this and all the following tables comes from studies on E12 rat embryos, while the right panel comes from studies on E13 rat embryos.

The somatic motor neurons are the first to be generated in the spinal gray (**Table 1A**, *left panel*), starting out in the 3.2- to 4.5-mm specimens (Volume 1, 2nd ed.). At the cervical level, peak generation occurs from 6.3- to 8.0-mm specimens, when the ventral spinal neuroepithelium (NEP) appears to be in a stockbuilding stage (**Fig 1A**) but most stem cells are actually producing postmitotic neurons. Thus, the NEPs having their final neurogenetic divisions appear most active and robust at that time. But shortly thereafter, postmitotic neurons gather in the basal parts of the NEP (**Fig. 1B**) and their cell bodies accumulate more and more cytoplasm. Only a short time later motor neurons reach their peak generation at all lower levels (**Table 1A**, *right panel*). In rats, we also found that contra-

laterally projecting sensory relay neurons are generated earlier than the ipsilaterally projecting ones (Altman and Bayer, 1984). Human spinal cords have the same neurons and the same developmental sequence probably exists. The earlier generation of the contralateral neurons is most likely related to the fact that these neurons have longer axons.

In the medulla, the oldest motor neurons are in the hypoglossal nucleus with an estimated peak in 6.3- to 8.0-mm specimens. By the time that specimens have reached 10- to 10.5-mm, there are no longer any stem cells for hypoglossal motor neurons in the medullary NEP (note the blank entry in **Table 1A**, *right panel*). The dorsal motor nucleus of the vagus follows slightly behind, because some neurogenesis occurs in CR 10- to 10.5-mm specimens. The motor neurons shown in Figure 2B may be destined to settle in the hypoglossal nucleus.

Sensory neurons in gracilis, cuneatus, and solitary nuclei begin neurogenesis by 6.3-8 mm and have robust neurogenesis by 10-10.5 mm. All of these neurons form prominent fiber tracts that relay sensory information to the midbrain and eventually to the thalamus. But the NEPs generating these neurons in the dorsal medulla appear as stockbuilding because the peak days of neurogenesis give no indication that neurons are actually being generated (**Figs. 2A, 2B**).

The vestibular nuclei, especially the lateral nucleus are being generated in the specimens presented here. These nuclei are closely related to the cerebellum and help to maintain balance. They, along with the hypoglossal prepositus nucleus, are also involved in regulating eye movements. In contrast, the audi-tory nuclei in the medulla (cochlear nuclei) are just starting out, so the NEPs producing these structures are mainly in stockbuilding stages.

The reticular formation and the raphe complex are beginning generation in the 6.3- to 8-mm specimens and continuing more robustly in the 10- to 10.5-mm specimens. The postmitotic raphe neurons are laterally placed. Medial raphe neurons will be generated later.

THE SPINAL CORD VENTRAL HORN IN
METHACRYLATE-EMBEDDED RAT EMBRYOS

A. Embryonic day 12

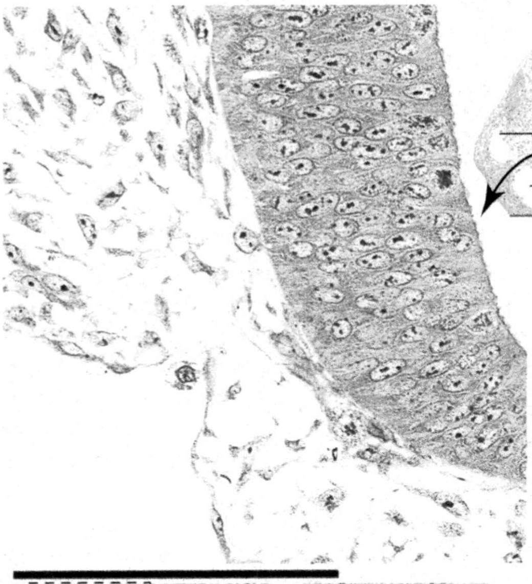

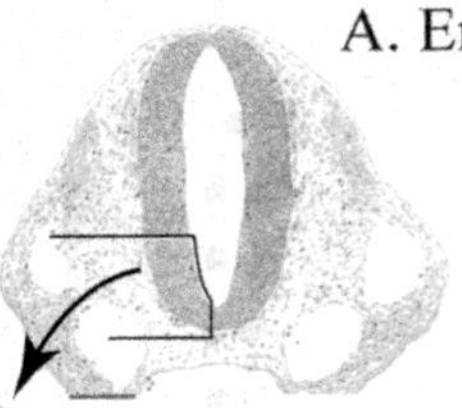

Neurogenesis is active in the ventral horn (**Table 1A**), especially at the cervical level (shown here). Many stem cells are producing postmitotic neurons in their neurogenetic phase. But this neuroepithelium (NEP) has the same appearance as a stockbuilding NEP. It is only because of our past research using ³H-Thymidine autoradiography that we know the dynamics of neuron production by NEP stem cells. Stem cells were already producing neurons in specimens with CRs from 3.2 to 4.5 mm (Volume 1, 2ⁿᵈ Ed.).

B. Embryonic day 13

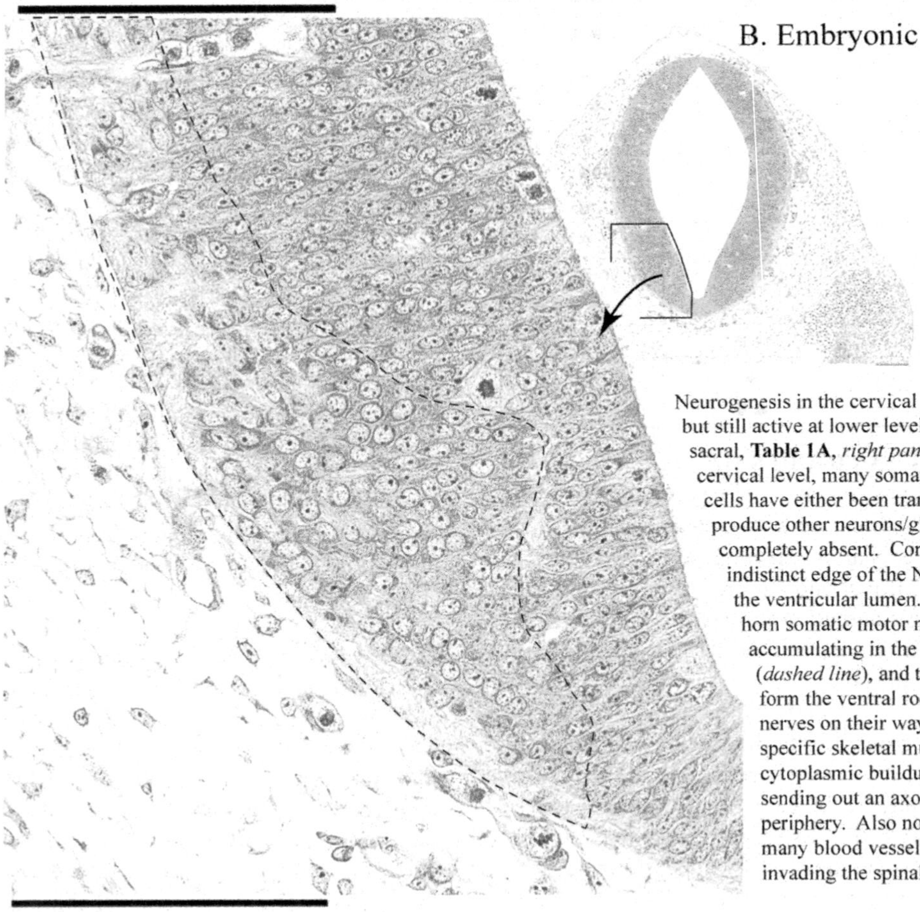

Neurogenesis in the cervical level is waning but still active at lower levels (thoracic to sacral, **Table 1A**, *right panel*). But at this cervical level, many somatic motor stem cells have either been transformed to produce other neurons/glia or are completely absent. Consequently, the indistinct edge of the NEP is closer to the ventricular lumen. The ventral horn somatic motor neurons are accumulating in the basal NEP (*dashed line*), and their axons will form the ventral roots of the spinal nerves on their way to innervate specific skeletal muscles. Note the cytoplasmic buildup in advance of sending out an axon into the periphery. Also notable are the many blood vessels that are invading the spinal cord wall.

Figure 1. The spinal cord in rat embryos at similar stages as human specimens with CR lengths of 6.3-8.0 mm (**A, E12**) and 10.0-10.5 mm (**B, E13**). (3µ methacrylate section, cresyl violet stain) scale bars=0.25 mm.
braindevelopmentmaps.org (E12 and E13 coronal archives)

Figure 2. The medulla in rat embryos at similar stages as human specimens with CR lengths of 6.3-8.0 mm (**A, E12**) and 10.0-10.5 mm (**B, E13**). (3µ methacrylate sections, cresyl violet stain) scale bars=0.25 mm.
braindevelopmentmaps.org (E12 and E13 coronal archives)

THE MEDULLA IN METHACRLATE-EMBEDDED RAT EMBRYOS

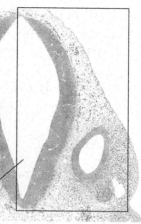

A. Embryonic day 12

The neuroepithelium (NEP) occupies entire brain wall even though neurogenesis is active in many populations. Most NEPs are still in the stockbuilding stage, adding more stem cells prior to their neurogenetic divisions.

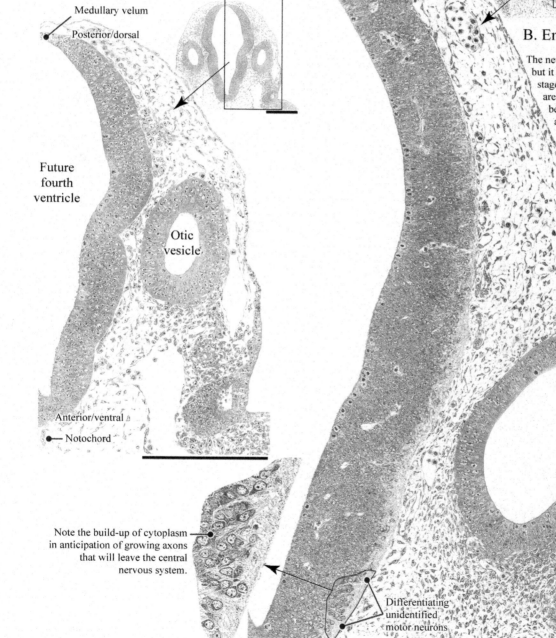

Medullary velum

Posterior/dorsal

Future fourth ventricle

Otic vesicle

Anterior/ventral

Notochord

B. Embryonic day 13

The neuroepithelium is still thick but it is now in its neurogenetic stage. The many cells at its base are postmitotic neurons that are beginning to migrate outward and differentiate, especially ventrally.

Note the build-up of cytoplasm in anticipation of growing axons that will leave the central nervous system.

Differentiating unidentified motor neurons

Floor plate

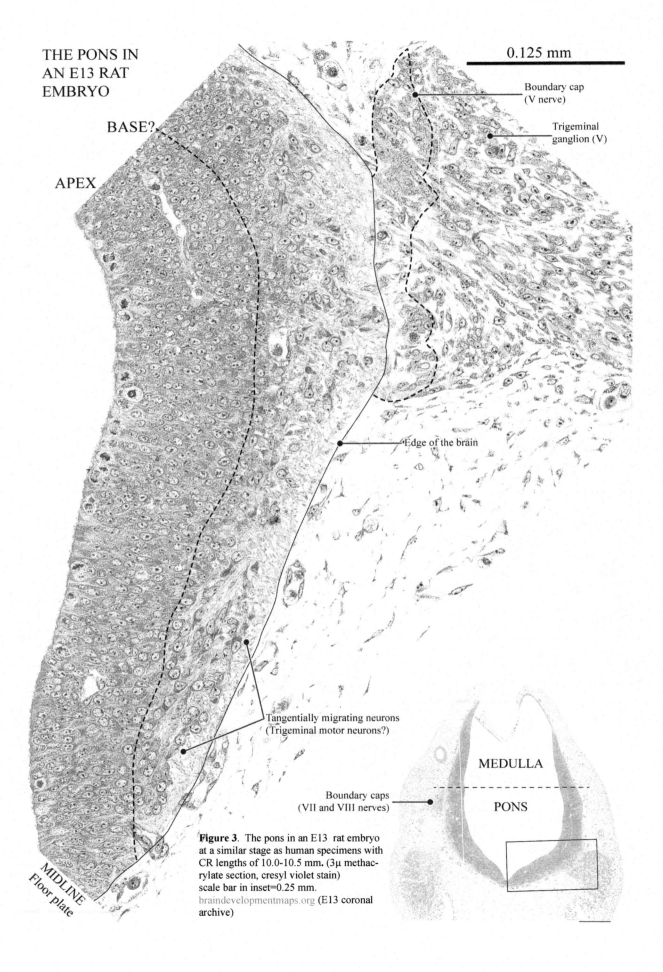

THE PONS IN
AN E13 RAT
EMBRYO

BASE?

APEX

0.125 mm

Boundary cap
(V nerve)

Trigeminal
ganglion (V)

Edge of the brain

Tangentially migrating neurons
(Trigeminal motor neurons?)

MIDLINE
Floor plate

Boundary caps
(VII and VIII nerves)

MEDULLA

PONS

Figure 3. The pons in an E13 rat embryo at a similar stage as human specimens with CR lengths of 10.0-10.5 mm. (3µ methacrylate section, cresyl violet stain) scale bar in inset=0.25 mm.

braindevelopmentmaps.org (E13 coronal archive)

TABLE 1B/FIGURES 3-4

Many of the differentiating neurons outside the pontine NEP on E13 (**Fig. 3**) were generated on E12 in rats, and we presume the same is true in human specimens. At this anterior level, the likely suspects are the tangentially migrating trigeminal motor neurons, which have finished generation on E12 (*left column*, **Table 1B**); no progenitors remain in the NEP on E13 (*blank space, right column*, **Table 1B**). Other neurons outside the indistinct basal edge of the pontine NEP might be neurons in the locus coeruleus, medial trapezoid nucleus, and pioneer neurons in the reticular formation.

The E13 rat cerebellum (equivalent to human specimens with CRs of 10 to 10.5 mm) is still very immature (**Table 1B**, *right column*, **Fig. 4**) with many mitotic cells at the NEP apex. These progenitors are just beginning to generate some deep nuclear neurons and Purkinje cells, so none would be outside the NEP in any differentiating zones. The precerebellar nuclei, especially the inferior olive are being generated in the medulla (**Table 1B**). Since E13 (10.0- to 10.5-mm human embryos) is a peak day for the inferior olive, it is likely that their neurons are still in the medullary NEP. The medullary NEP is very active at this time, only a few identifications can be made, and those are uncertain (**Fig. 2A, 2B**).

Table 1B: Neurogenesis by Region

REGION and NEURAL POPULATION	CROWN RUMP LENGTH 6.3-8.0 mm	10.0-10.5 mm
PONS		
Trigeminal (V) principal sensory		•
Infratrigeminal	•	•
Trigeminal (V) motor	• •	
Facial (VII) nucleus	•	• •
Locus coeruleus	• •	•
Parabrachial		•
Dorsal tegmental	•	• •
Raphe complex		• •
Reticular formation	•	• •
Dorsal n. lateral lemniscus	•	• •
Ventral n. lateral lemniscus		•
Lateral trapezoid nucleus	• •	
Medial superior olive	•	• •
Lateral superior olive	•	•
CEREBELLUM and PRECEREBELLAR NUCLEI		
Deep nuclei		•
Purkinje cells		•
External cuneate	•	•
Inferior olive		• •
Lateral reticular		•

THE CEREBELLUM IN AN E13 RAT EMBRYO

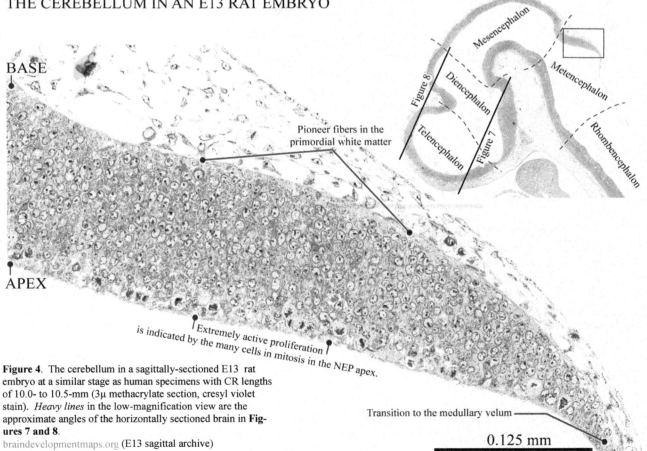
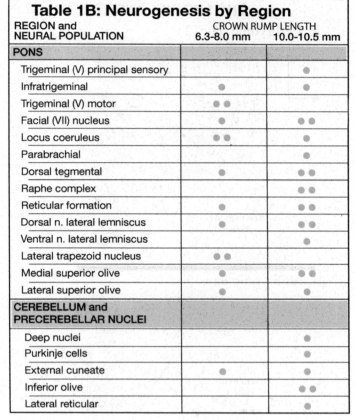

Pioneer fibers in the primordial white matter

BASE

APEX

Extremely active proliferation is indicated by the many cells in mitosis in the NEP apex.

Transition to the medullary velum

0.125 mm

Figure 4. The cerebellum in a sagittally-sectioned E13 rat embryo at a similar stage as human specimens with CR lengths of 10.0- to 10.5-mm (3μ methacrylate section, cresyl violet stain). *Heavy lines* in the low-magnification view are the approximate angles of the horizontally sectioned brain in **Figures 7 and 8**.

braindevelopmentmaps.org (E13 sagittal archive)

THE MIDBRAIN TEGMENTUM IN RAT EMBRYOS

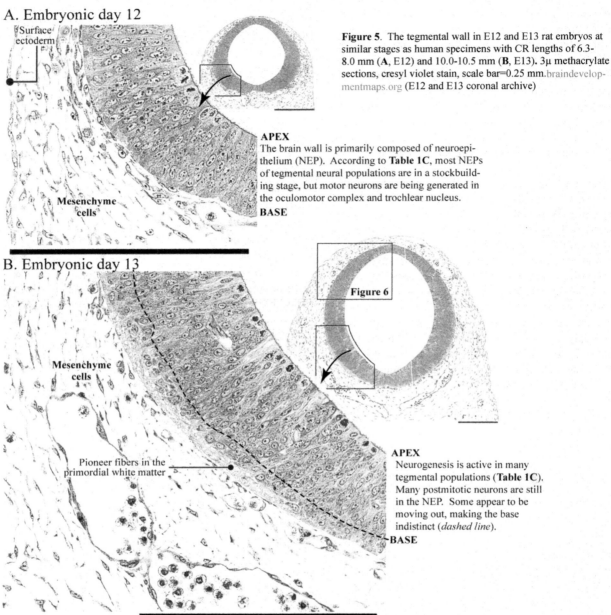

A. Embryonic day 12

Surface ectoderm

Mesenchyme cells

Figure 5. The tegmental wall in E12 and E13 rat embryos at similar stages as human specimens with CR lengths of 6.3-8.0 mm (**A**, E12) and 10.0-10.5 mm (**B**, E13). 3μ methacrylate sections, cresyl violet stain, scale bar=0.25 mm. braindevelopmentmaps.org (E12 and E13 coronal archive)

APEX
The brain wall is primarily composed of neuroepithelium (NEP). According to **Table 1C**, most NEPs of tegmental neural populations are in a stockbuilding stage, but motor neurons are being generated in the oculomotor complex and trochlear nucleus.
BASE

B. Embryonic day 13

Mesenchyme cells

Pioneer fibers in the primordial white matter

Figure 6

APEX
Neurogenesis is active in many tegmental populations (**Table 1C**). Many postmitotic neurons are still in the NEP. Some appear to be moving out, making the base indistinct (*dashed line*).
BASE

TABLE 1C/FIGURES 5-6

The midbrain tegmentum (**Fig. 5**) and the tectum (**Fig. 6**) show high-resolution images of methacrylate-embedded rat embryos on E12, comparable to the CR 6.3- to 8.0-mm specimens, and E13, comparable to the CR 10- to 10.5-mm specimens in this volume. The oldest neurons in the tegmentum are the primary sensory neurons in the mesencephalic nucleus (**Table 1C**, *left column)*, but these are generated outside the brain, and they have not yet migrated in (judging from the photo in **Fig. 3** that shows the trigeminal nerve entry zone in the pons).

The few differentiating neurons outside the base of the tegmental NEP on E13 (*dashed line*, **Fig. 5B**) were gen-erated on E12 in rats, and we presume the same is true in human specimens. Although E12 is a peak day of generation for motor nuclei in the mesencephalon (oculomotor and trochlear, **Table 1C**, *left column*), the E12 NEP gives every appearance of a typical stockbuilding active proliferation stage. As we have said earlier, the peak days of generation are the days when the NEPs are most active and very thick. At the level pictured in **Figure 5**, we may be cutting through the NEP that will generate the somatic motor neurons in the oculomotor nucleus and some of the cells outside the NEP on E13 may be differentiating neurons destined to settle there. But the NEP is thicker on E13 than it is on E12, and shows a high level of proliferation, even though all oculomotor neurons have been gen-

Table 1C: Neurogenesis by Region

REGION and NEURAL POPULATION	CROWN RUMP LENGTH	
	6.3-8.0 mm	10.0-10.5 mm
MESENCEPHALIC TEGMENTUM/ISTHMUS		
Mesencephalic V nucleus	•	
Oculomotor III nucleus	• •	
Edinger Westphal III nucleus	•	• •
Trochlear IV nucleus	• •	
Nucleus of Darkschewitsch	• •	• •
Parabigeminal nucleus		•
Raphe complex	• •	• •
Dorsal central gray		•
Lateral central gray		•
Ventral central gray		• •
Red nucleus (magnocellular)		• •
Red nucleus (parvocellular)		• •
Substantia nigra compacta		• •
Substantia nigra reticulata		• •
Ventral tegmental area (lateral)		•
Dorsal interpeduncular nucleus		•
Ventral interpeduncular nucleus		• •
MESENCEPHALIC TECTUM		
Superior colliculus magnocellular zone	•	• •
Superior colliculus stratum profundum		•
Superior colliculus stratum album		•
Superior colliculus stratum lemnisci		• •

erated. This section may also be cutting through the NEP that is generating the red nucleus (**Table 1C**, *right column*) and E13 is the day of its peak generation in both magnocellular and parvocellular parts. On the other hand, this section may be cutting through the NEPs that will generate the substantia nigra and ventral tegmental area. After close examination of all the sections through the mesencephalon, none of them appear that much different from each other, so these sections were chosen because they are at a middle anterior/posterior level. We use the frontal/horizontal plane because those sections cut through the NEP nearly perpendicular to its longitudinal axis.

The tectum (**Fig. 6**) is still quite immature, with very few cells outside the NEP on E13 in rats. Again, we assume that human specimens with CR lengths of 10-10.5 mm are at a similar state of immaturity. The most likely suspects for identification of the few cells outside the tectal NEP are the superior colliculus neurons that will settle in the magnocellular zone, because some of them were generated on E12 in rats (**Table 1C**, *left column*), and their peak time of generation is on E13.

THE MIDBRAIN TECTUM IN AN E13 RAT EMBRYO

APEX

Postmitotic neuron

BASE

Postmitotic neuron

Figure 6. The tectum in an E13 rat embryo at a similar stage as human specimens with CR lengths of 10.0-10.5 mm. 3µ methacrylate section, cresyl violet stain. braindevelopment-maps.org (E13 coronal archive)

The neuroepithelium is actively proliferating and generating postmitotic neurons because this is the the peak time of generation for several populations in the superior colliculus (**Table 1C**, *right column*). The few neurons outside the base of the NEP were generated on E12, possibly those destined to settle in the magnocellular zone.

0.125 mm

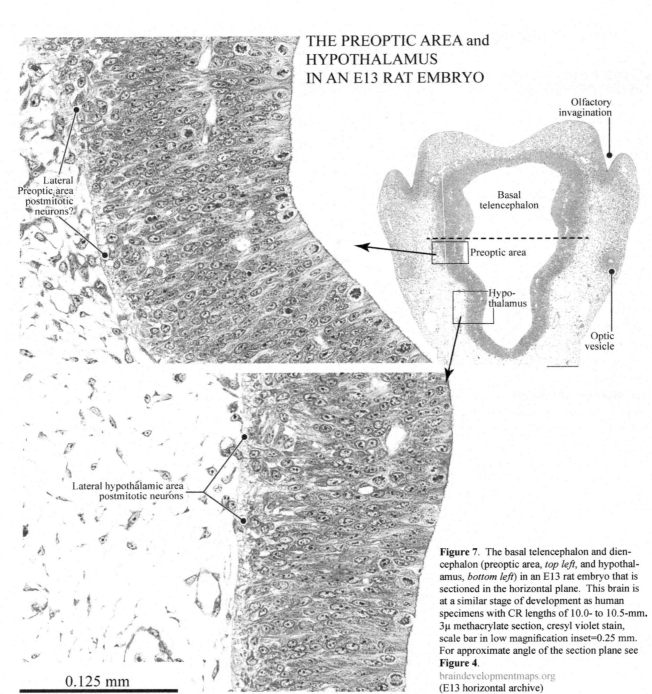

THE PREOPTIC AREA and
HYPOTHALAMUS
IN AN E13 RAT EMBRYO

Figure 7. The basal telencephalon and diencephalon (preoptic area, *top left*, and hypothalamus, *bottom left*) in an E13 rat embryo that is
sectioned in the horizontal plane. This brain is
at a similar stage of development as human
specimens with CR lengths of 10.0- to 10.5-mm.
3μ methacrylate section, cresyl violet stain,
scale bar in low magnification inset=0.25 mm.
For approximate angle of the section plane see
Figure 4.
braindevelopmentmaps.org
(E13 horizontal archive)

TABLE 1D/FIGURES 7-8

The preoptic area, hypothalamus (**Fig. 7**) and the thal-
amus/epithalamus (**Fig. 8**) show high-resolution images
of horizontally sectioned, methacrylate-embedded rat
embryos on E13, comparable to the CR 10- to 10.5-mm
specimens in this volume. The oldest neurons in the hypothalamus are in the lateral preoptic and lateral hypothalamic areas (**Table 1D**, *left column)*. These areas stretch
along the anterior/posterior extent of the hypothalamus
with no distinction between them. Both of these populations started neurogenesis on E12 in rats and E13 is their
peak day of origin (**Table 1D**, *right column*). It seems safe
to conclude that the cells that are horizontally oriented

outside the base of the NEP are pioneer neurons in these
populations in both the preoptic area (**Fig 7**, *top*) and the
hypothalamus proper (**Fig. 7**. *bottom*). The lateral mammillary nucleus is another population of older neurons,
but the mammillary body NEP is not in the plane of this
section.

Only three neural populations are beginning to be
generated in the thalamus/epithalamus on E13 (**Table
1D**, *right column*). In the horizontal section shown in
Figure 8, the thalamic neuroepithelium is very thick with
a feathered basal edge. The area outlined by a dashed
line very tentatively may enclose some postmitotic neu-

rons that will settle in the oldest populations (medial geniculate, reticular, and lateral habenular nuclei).

But the neuroepithelial continuum pictured in **Figures 7-8** is busy generating a variety of other neuronal populations on E13. There is little to no demarcation between a NEP that may be generating the supraoptic nucleus and another NEP that may be generating the paraventricular nucleus in the hypothalamus, for example. In a Nissl-stained section, where all parts of the NEP appear to be the same, we cannot visualize the finer details of the NEP mosaic. What we can say with certainty is that the high-magnification NEPs shown in **Figures 7-8** are in the preoptic area, hypothalamus, and thalamus, respectively. Hopefully, markers for differential gene expression will reveal more detail about the specific identity of the many different neural progenitors in these neuroepithelia.

Table 1D: Neurogenesis by Region		
REGION and NEURAL POPULATION	CROWN RUMP LENGTH 6.3-8.0 mm	10.0-10.5 mm
HYPOTHALAMUS		
Lateral preoptic area	●	●●
Lateral hypothalamic area	●	●●
Medial preoptic area		●
Median preoptic nucleus		●
Anterobasal nucleus		●●
Supraoptic nucleus		●●
Paraventricular nucleus		●
Ventromedial nucleus		●
Premammillary nucleus		●
Lateral mammillary nucleus	●	●●
THALAMUS and EPITHALAMUS		
Medial geniculate nucleus		●
Reticular nucleus		●
Lateral habenular nucleus		●

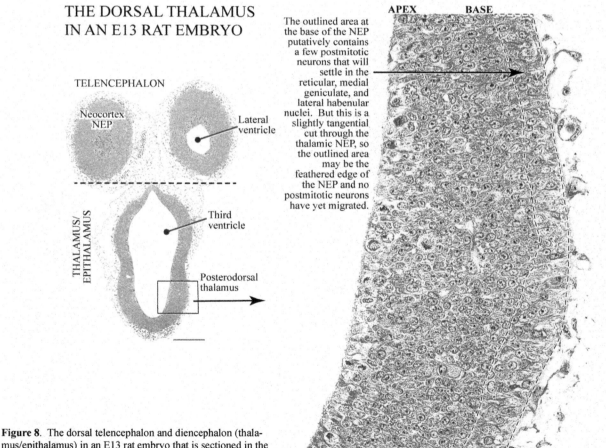

THE DORSAL THALAMUS IN AN E13 RAT EMBRYO

TELENCEPHALON

Neocortex NEP

Lateral ventricle

THALAMUS/ EPITHALAMUS

Third ventricle

Posterodorsal thalamus

APEX   BASE

The outlined area at the base of the NEP putatively contains a few postmitotic neurons that will settle in the reticular, medial geniculate, and lateral habenular nuclei. But this is a slightly tangential cut through the thalamic NEP, so the outlined area may be the feathered edge of the NEP and no postmitotic neurons have yet migrated.

0.125 mm

Figure 8. The dorsal telencephalon and diencephalon (thalamus/epithalamus) in an E13 rat embryo that is sectioned in the horizontal plane. This brain is at a similar stage of development as human specimens with CR lengths of 10.0- to 10.5-mm. 3μ methacrylate section, cresyl violet stain, scale bar in low-magnification inset=0.25 mm. For approximate angle of the section plane see **Figure 4**.

12

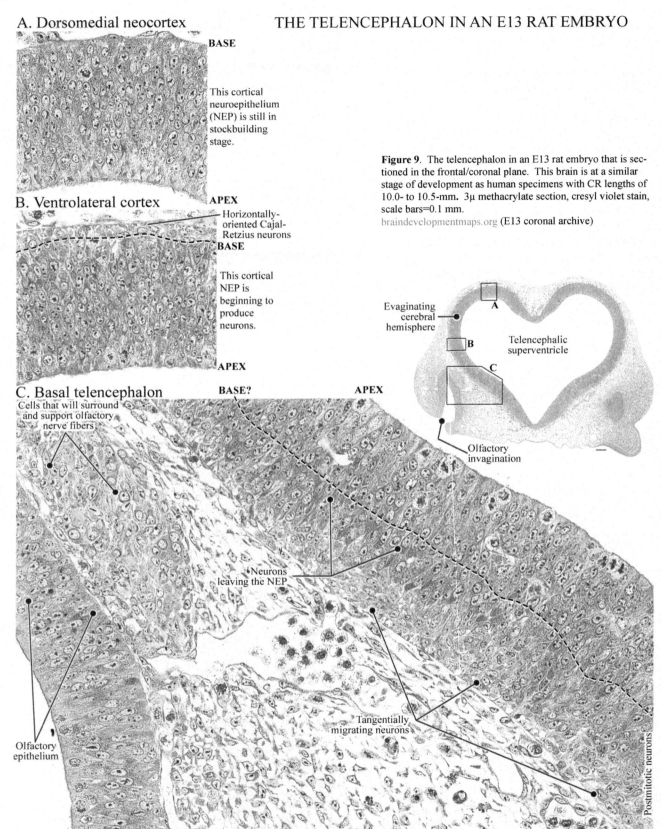

A. Dorsomedial neocortex

BASE

This cortical neuroepithelium (NEP) is still in stockbuilding stage.

APEX

B. Ventrolateral cortex

APEX

Horizontally-oriented Cajal-Retzius neurons

BASE

This cortical NEP is beginning to produce neurons.

APEX

Figure 9. The telencephalon in an E13 rat embryo that is sectioned in the frontal/coronal plane. This brain is at a similar stage of development as human specimens with CR lengths of 10.0- to 10.5-mm. 3μ methacrylate section, cresyl violet stain, scale bars=0.1 mm.

braindevelopmentmaps.org (E13 coronal archive)

Evaginating cerebral hemisphere

A

B

C

Telencephalic superventricle

Olfactory invagination

C. Basal telencephalon

BASE?   APEX

Cells that will surround and support olfactory nerve fibers

Neurons leaving the NEP

Tangentially migrating neurons

Olfactory epithelium

Postmitotic neurons

TABLE 1E/FIGURE 9

Except for the lateral basal telencephalon and the amygdala, the neuroepithelia in various parts of the telencephalon are still mainly in the stock-building stage (**Table 1E**). The section shown in **Figure 9** cuts through the basal telencephalon and the evaginating cerebral cortex; the amygdala is not in the section plane. The lateral basal telencephalon has many postmitotic neurons outside the base of the NEP, making the edge indistinct (**Fig. 9C**, *dashed line*). The postmitotic neurons can be divided into two groups: a subpial group has many tangentially migrating neurons interspersed among some blank white spaces that may be glial channels. A lot of migration occurs in this part of the brain because the output neurons that will settle in the olfactory bulb are generated before there is any olfactory bulb evagination from the basal forebrain (especially the output neurons of the accessory bulb where E13 is the peak day of their generation, **Table 1E**, *right panel*). A few mitral neurons are also being generated by NEPs in the basal telencephalon on E13 and will migrate forward into the olfactory bulb after it evaginates. The entopeduncular nucleus is a group of the oldest neurons in the basal telencephalon, but they cannot be distinguished from the other postmitotic neurons outside the NEP.

In the periphery, a well-defined clump of cells will surround and support the axons of the primary sensory neurons in the olfactory epithelium (**Fig. 9C**). But at this time, there is no evidence that olfactory nerve fibers are being produced. This clump of cells may have something to do with inducing the differentiation of the olfactory epithelium from the surrounding nasal epithelium.

In the future ventrolateral cerebral cortex (**Fig. 9B**), neurogenesis is just beginning, and a few horizontally oriented neurons outside the NEP can be tentatively identified as Cajal-Retzius neurons (**Table 1E**, *right column*). This part of the cortex matures earlier than the dorsomedial part (**Fig. 9A**), where there is no evidence that any postmitotic neurons are outside the cortical NEP. Thus even at this earliest stage, a maturation gradient is evident. The cortex in the ventrolateral part will most likely develop into the insular area, part of the lateral limbic cortex. This is where the first evidence of a cortical plate will appear. On the other hand, the dorsomedial cortex will most likely develop into neocortex.

Table 1E: Neurogenesis by Region

REGION and NEURAL POPULATION	CROWN RUMP LENGTH 6.3-8.0 mm	10.0-10.5 mm
PALLIDUM AND STRIATUM		
Entopeduncular nucleus	●	● ●
Globus pallidus (external segment)		● ●
Substantia innominata		●
Basal nucleus of Meynert		●
Olfactory tubercle (large neurons)		●
AMYGDALA		
Anterior amygdaloid area		●
Nucleus of the accessory olfactory tract	●	● ●
Central nucleus		●
Medial nucleus		●
Anterior cortical nucleus		● ●
Bed n. stria terminalis (anterior)		●
Bed n. stria terminalis (preoptic continuation)		●
SEPTUM		
Medial nucleus		●
Diagonal band (vertical limb)		●
OLFACTORY CORTEX		
Layers III-IV (anterior)		●
Layers III-IV (posterior)		●
OLFACTORY BULB		
Output neurons (accessory bulb)	●	● ●
Mitral cells (main bulb)		●
Internal tufted cells (main bulb)		●
NEOCORTEX		
Cajal-Retzius neurons		●

REFERENCES

Altman J, Bayer SA (1984) *The Development of the Rat Spinal Cord*. Advances in Anatomy Embryology and Cell Biology, Vol. 85, Berlin, Springer -Verlag.

Bayer SA, Altman J, Russo RJ, Zhang X (1993) Timetables of neurogenesis in the human brain based on experimentally determined patterns in the rat. *Neurotoxicology* **14**: 83-144.

Bayer SA, Altman J, Russo RJ, Zhang X (1995) Embryology. In: *Pediatric Neuropathology*, Serge Duckett, Ed. Williams and Wilkins, pp. 54-107.

Bayer SA, Altman J (1995) Development: Some principles of neurogenesis, neuronal migration and neural circuit formation. In: *The Rat Nervous System*, 2nd Edition, George Paxinos, Ed. Academic Press, Orlando, Florida., pp. 1079-1098.

Bayer SA, Altman J (2008) *Atlas of Human Central Nervous System Development* (First Edition), Volume 5, CRC Press.

Bayer SA, Altman J (2012-present) www.neurondevelopment.org (This website has downloadable pdf files of our scientific papers on rat brain development grouped by subject.)

Bayer, SA (2013-present) www.braindevelopmentmaps.org (This website is a database containing methacrylate-embedded normal rat embryos and paraffin-embedded rat embryos exposed to ^{3}H-Thymidine.)

Glossary entries are based on my knowledge of embryology, neuroanatomy, neurophysiology, and Altman and Bayer's previous research on nervous system development. Nearly all entries were also checked with searches on Google, PubMed, PubMed Stat Pearls, and Wikipedia.

Loughna P, Citty L, Evans T, Chudleigh T (2009) Fetal size and dating: Charts recommended for clinical obstetric practice, *Ultrasound*, 17:161-167.

O'Rahilly R; Müller F. (1987) *Developmental Stages in Human Embryos, Carnegie Institution of Washington*, Publication 637.

O'Rahilly R; Müller F. (1994) *The Embryonic Human Brain*, Wiley-Liss, New York.

PART II: M2300
CR 6.3 mm (GW 5.0)
Frontal/Horizontal

This specimen is embryo #2300 in the Minot Collection, designated here as M2300. The crown-rump length (CR) is 6.3 mm estimated to be at gestational week (GW) 5.0. M2300's prosencephalic and anterior mesencephalic sections are cut (8 μm) in the coronal plane, but the plane shifts to predominantly horizontal in the posterior mesencephalon, pons, and medulla. We photographed 48 sections at low magnification from the frontal prominence to the posterior tips of the mesencephalon and medulla. Fourteen of these sections are illustrated in **Plates 1AB to 13AB**. All photographs were used to generate computer-aided 3-D reconstructions of the external features of M2300's brain and optic vesicle (**Figure 10**), and to show each illustrated section *in situ* (*insets*, **Plates 1A to 13A**). Each illustrated section shows the brain with all surrounding tissues. Labels in **A Plates** (normal-contrast images) identify non-neural and peripheral neural structures; labels in **B Plates** (low-contrast images) identify central neural structures.

The forebrain is the smallest brain vesicle with a mainly stockbuilding neuroepithelium in the telencephalon surrounding a small prosencephalic superventricle. Anterior sections are tentatively identified as the future telencephalon, while sections through the optic vesicle and posterior to it are more clearly identified as diencephalic. The diencephalic NEP, especially in hypothalamic areas, are just beginning to generate neurons in the lateral preoptic and hypothalamic areas, and in the mammillary nucleus so it is no longer purely stockbuilding. These NEPs are starting the neurogenetic phase but they are still very thick and enlarging. Cell migration is absent in the prosencephalon and diencephalon. The olfactory placode is a prominent feature in the anterolateral head. The optic vesicle is just beginning to curve around a definite lens placode. The optic vesicle has an outer retinal neuroepithelium (thick with presumptive glial channels adjacent to the lens placode) and a pigment epithelium (thin) looping above and behind the future retina.

The NEP in the mesencephalic tectum and the cerebellar primordium are exclusively stockbuilding. Tegmental and isthmal NEPs, pontine NEPs, medullary NEPs and ventral spinal NEPs are generating motor neurons and other neurons like reticular formation populations, the locus coeruleus, and a few raphe neurons. Even though these NEPs are well into the neurogenetic stage, they are still expanding.

From the subthalamic/lateral hypothalamic area, through the mesencephalic tegmentum, the pons, the medulla, and the spinal cord there is a *prominent accumulation of fibers beneath the pia*. This starts in **Plate 4** and goes all the way to the medullary part of **Plate 13**. It is unlikely that these axons are from centrally generated neurons because so few are postmitotic at this time. However, peripheral ganglia are recognizable in the head and the spinal cord and are generally well developed. We postulate that many of these subpial fibers are from the ingrowing cranial and spinal nerves. These nerves appear to invade the basal parts of the NEPs and may have something to do with specifying newly generated neurons accumulating in the basal NEP before migration. Many neurons sequester there for hours (rats) or days (humans). Are they waiting for the right fibers to grow past and initiate their journey to a settling site?

Rhombomeres are still prominent in the NEPs of the pons and medulla. These NEPs are all in neurogenetic phase. Many sensory neurons are being generated in each rhombomere at this time, especially in R4 and R5, the medial superior olive and the lateral trapezoid nucleus are in production. In R3, R6, and R7, neurons destined to settle in the solitary nucleus are just beginning to be generated. Some of the feathered edges of the NEPs in these rhombomeres are tentatively identified as postmitotic sensory neurons in the following plates.

M2300 Computer-aided 3-D Brain Reconstructions

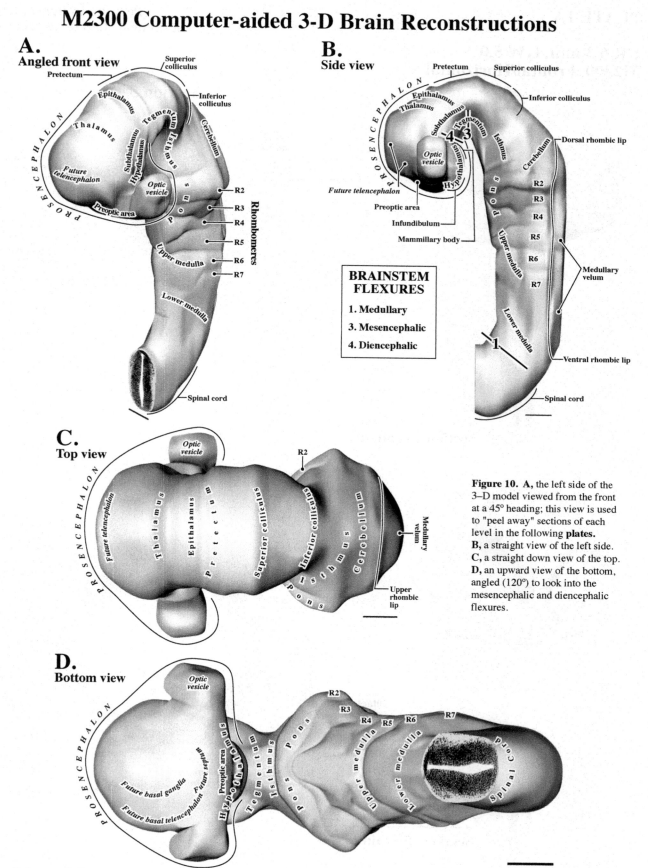

A. Angled front view

Pretectum
Superior colliculus
Epithalamus
Inferior colliculus
Thalamus
Tegmentum
Isthmus
Cerebellum
Subthalamus
PROSENCEPHALON
Hypothalamus
Future telencephalon
Optic vesicle
Pons
R2
R3
Preoptic area
R4
R5
Rhombomeres
Upper medulla
R6
R7
Lower medulla
Spinal cord

B. Side view

Pretectum
Superior colliculus
Epithalamus
Inferior colliculus
Thalamus
Subthalamus
Tegmentum
PROSENCEPHALON
4 3
Optic vesicle
Isthmus
Dorsal rhombic lip
Cerebellum
Hypothalamus
Future telencephalon
Pons
R2
Preoptic area
R3
Infundibulum
R4
Mammillary body
R5
Upper medulla
R6
R7
Medullary velum
Lower medulla
1
Ventral rhombic lip
Spinal cord

BRAINSTEM FLEXURES
1. Medullary
3. Mesencephalic
4. Diencephalic

C. Top view

Optic vesicle
R2
PROSENCEPHALON
Future telencephalon
Thalamus
Epithalamus
Pretectum
Superior colliculus
Inferior colliculus
Isthmus
Cerebellum
Pons
Medullary velum
Upper rhombic lip

Figure 10. A, the left side of the 3–D model viewed from the front at a 45° heading; this view is used to "peel away" sections of each level in the following **plates**. **B,** a straight view of the left side. **C,** a straight down view of the top. **D,** an upward view of the bottom, angled (120°) to look into the mesencephalic and diencephalic flexures.

D. Bottom view

Optic vesicle
R2
PROSENCEPHALON
R3
R4 R5 R6
R7
Preoptic area
Future basal ganglia
Future septum
Hypothalamus
Future basal telencephalon
Tegmentum
Isthmus
Pons
Upper medulla
Lower medulla
Spinal Cord

Scale bars = 0.5 mm

PLATE 1A

CR 6.3 mm, GW 5.0
M2300, Frontal/Horizontal

**Peripheral neural and
non-neural structures labeled**

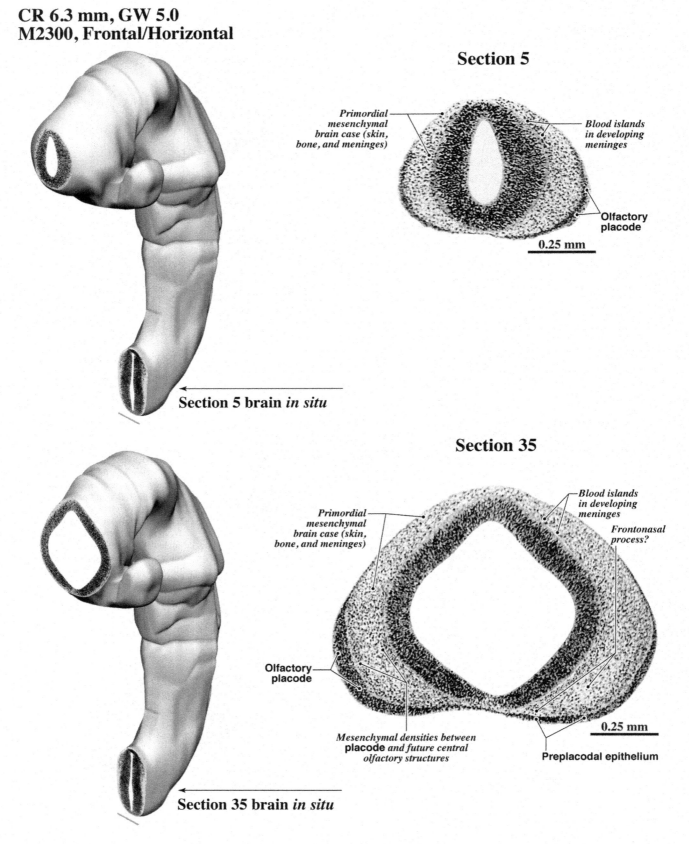

Section 5

Primordial mesenchymal brain case (skin, bone, and meninges)

Blood islands in developing meninges

Olfactory placode

0.25 mm

Section 5 brain *in situ*

Section 35

Primordial mesenchymal brain case (skin, bone, and meninges)

Blood islands in developing meninges

Frontonasal process?

Olfactory placode

Mesenchymal densities between **placode** *and future central olfactory structures*

Preplacodal epithelium

0.25 mm

Section 35 brain *in situ*

Section 5

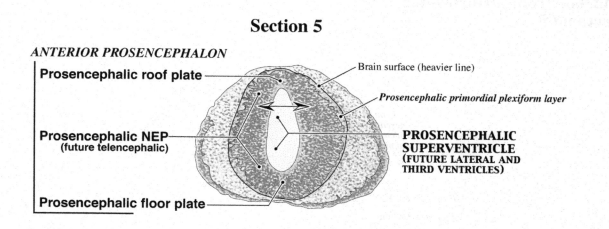

ANTERIOR PROSENCEPHALON

Prosencephalic roof plate

Brain surface (heavier line)

Prosencephalic primordial plexiform layer

Prosencephalic NEP
(future telencephalic)

PROSENCEPHALIC SUPERVENTRICLE (FUTURE LATERAL AND THIRD VENTRICLES)

Prosencephalic floor plate

Section 35

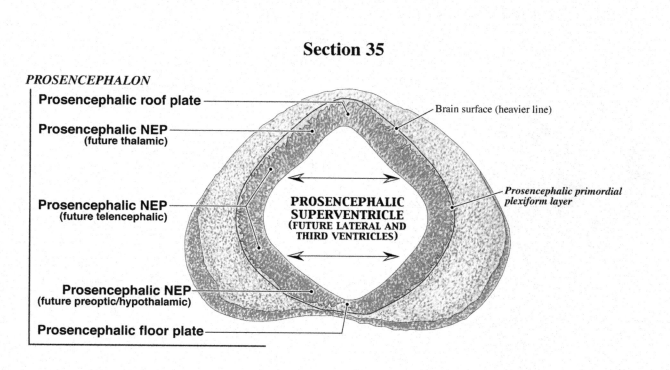

PROSENCEPHALON

Prosencephalic roof plate

Brain surface (heavier line)

Prosencephalic NEP
(future thalamic)

Prosencephalic NEP
(future telencephalic)

PROSENCEPHALIC SUPERVENTRICLE (FUTURE LATERAL AND THIRD VENTRICLES)

Prosencephalic primordial plexiform layer

Prosencephalic NEP
(future preoptic/hypothalamic)

Prosencephalic floor plate

NEP - Neuroepithelium

FONT KEY:
VENTRICULAR DIVISIONS - CAPITALS
Germinal zone - Helvetica bold
Transient structure - Times bold italic
Permanent structure - Times Roman or **Bold**

Arrows indicate the regionally *expanding shoreline* of the superventricle with increase in stockbuilding NEP cells.

PLATE 2A

CR 6.3 mm, GW 5.0
M2300, Frontal/Horizontal
Section 65

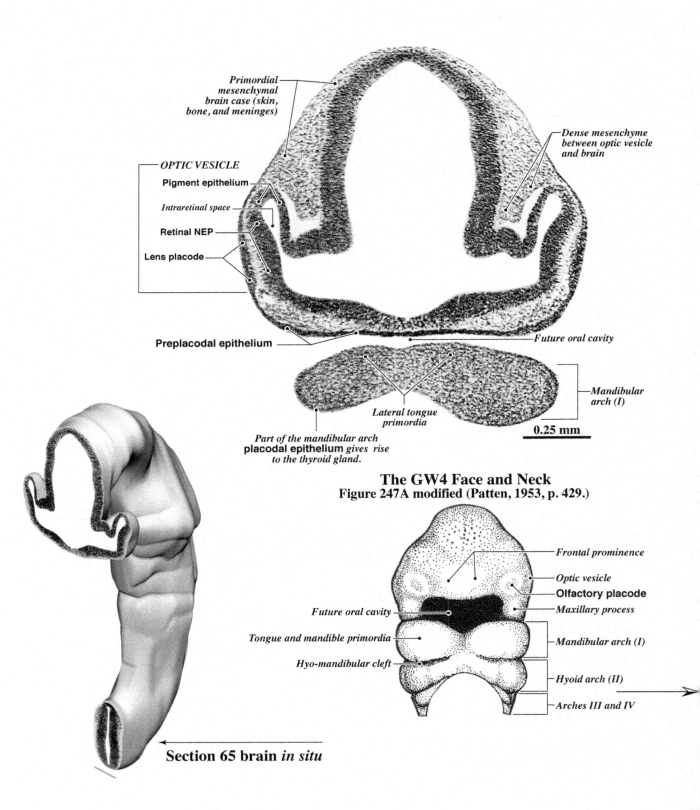

*Primordial
mesenchymal
brain case (skin,
bone, and meninges)*

*Dense mesenchyme
between optic vesicle
and brain*

OPTIC VESICLE

Pigment epithelium

Intraretinal space

Retinal NEP

Lens placode

Preplacodal epithelium

Future oral cavity

*Mandibular
arch (I)*

*Lateral tongue
primordia*

0.25 mm

Part of the mandibular arch
placodal epithelium *gives rise
to the thyroid gland.*

The GW4 Face and Neck
Figure 247A modified (Patten, 1953, p. 429.)

Frontal prominence

Optic vesicle

Olfactory placode

Maxillary process

Future oral cavity

Tongue and mandible primordia

Hyo-mandibular cleft

Mandibular arch (I)

Hyoid arch (II)

Arches III and IV

Section 65 brain *in situ*

DIENCEPHALON

THALAMUS

Diencephalic roof plate
(future choroid plexus in roof of third ventricle)

Brain surface (heavier line)

Thalamic NEP

Thalamic primordial plexiform layer

SUBTHALAMUS

Subthalamic NEP

DIENCEPHALIC SUPERVENTRICLE (FUTURE THIRD VENTRICLE)

Subthalamic primordial plexiform layer

Hypothalamic NEP

Glial channels in **retinal NEP**?

OPTIC RECESS

Diencephalic floor plate
(future chiasmal GEP)

Hypothalamic primordial plexiform layer

Chiasmal glial channels?

HYPOTHALAMUS

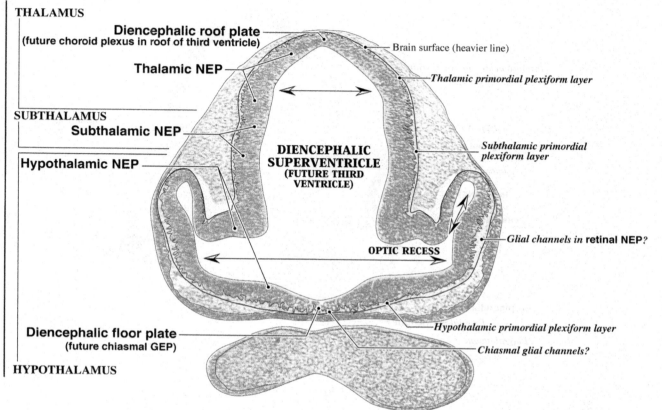

ABBREVIATIONS:
GEP - Glioepithelium
NEP - Neuroepithelium

Indicates expansion of the NEP by *stockbuilding* and *final neurogenetic divisions*.

FONT KEY:
VENTRICULAR DIVISIONS - CAPITALS
Germinal zone - Helvetica bold
Transient structure - Times bold italic
Permanent structure - Times Roman or **Bold**

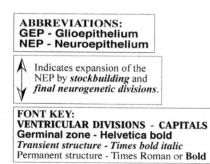

PLATE 3A

CR 6.3 mm, GW 5.0
M2300, Frontal/Horizontal
Section 75

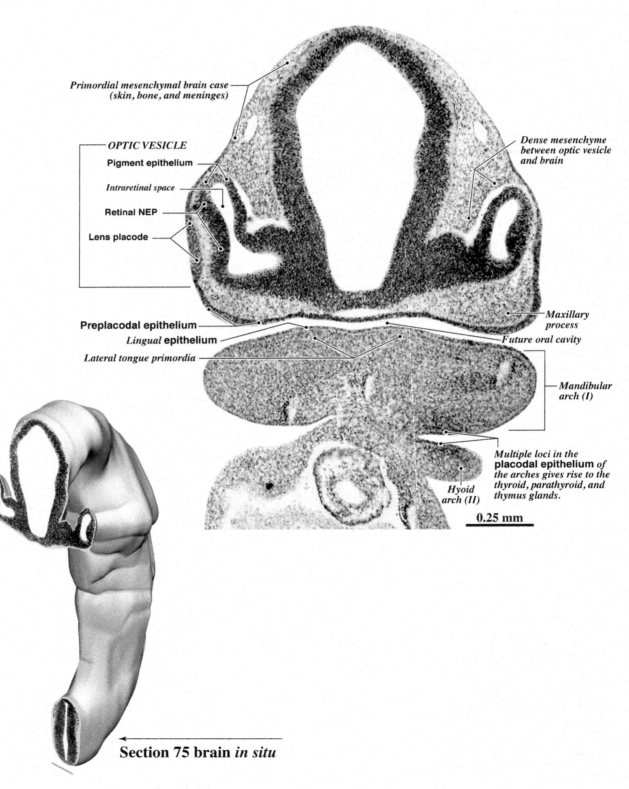

Primordial mesenchymal brain case
(skin, bone, and meninges)

Dense mesenchyme
between optic vesicle
and brain

OPTIC VESICLE

Pigment epithelium

Intraretinal space

Retinal NEP

Lens placode

Maxillary
process

Preplacodal epithelium

Lingual epithelium

Lateral tongue primordia

Future oral cavity

Mandibular
arch (I)

Multiple loci in the
placodal epithelium *of
the arches gives rise to the
thyroid, parathyroid, and
thymus glands.*

*Hyoid
arch (II)*

0.25 mm

Section 75 brain *in situ*

Central neural structures labeled

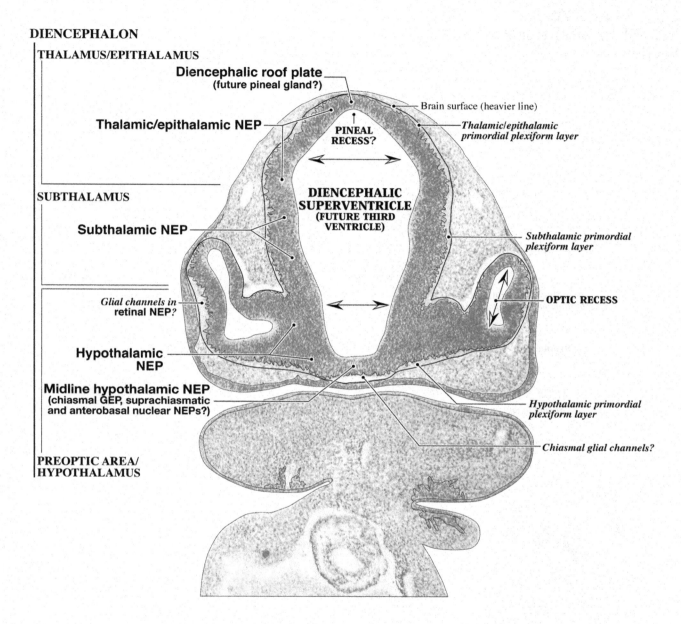

DIENCEPHALON

THALAMUS/EPITHALAMUS

Diencephalic roof plate
(future pineal gland?)

Brain surface (heavier line)

Thalamic/epithalamic NEP

*Thalamic/epithalamic
primordial plexiform layer*

**PINEAL
RECESS?**

**DIENCEPHALIC
SUPERVENTRICLE
(FUTURE THIRD
VENTRICLE)**

SUBTHALAMUS

Subthalamic NEP

*Subthalamic primordial
plexiform layer*

*Glial channels in
retinal NEP?*

OPTIC RECESS

**Hypothalamic
NEP**

Midline hypothalamic NEP
(chiasmal GEP, suprachiasmatic
and anterobasal nuclear NEPs?)

*Hypothalamic primordial
plexiform layer*

Chiasmal glial channels?

**PREOPTIC AREA/
HYPOTHALAMUS**

ABBREVIATIONS:
GEP - Glioepithelium
NEP - Neuroepithelium

 Indicates expansion of the
NEP by *stockbuilding* and
final neurogenetic divisions.

FONT KEY:
VENTRICULAR DIVISIONS - CAPITALS
Germinal zone - Helvetica bold
Transient structure - Times bold italic
Permanent structure - Times Roman or **Bold**

PLATE 4A

CR 6.3 mm, GW 5.0
M2300, Frontal/Horizontal
Section 85

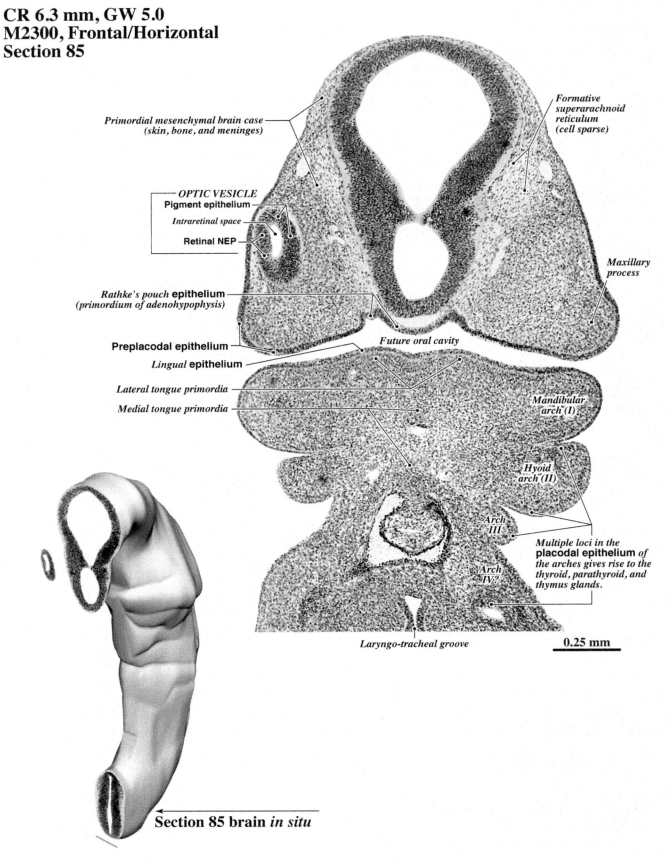

Primordial mesenchymal brain case
(skin, bone, and meninges)

Formative
superarachnoid
reticulum
(cell sparse)

OPTIC VESICLE
Pigment epithelium
Intraretinal space
Retinal NEP

Maxillary
process

Rathke's pouch **epithelium**
(primordium of adenohypophysis)

Preplacodal **epithelium**

Lingual **epithelium**

Future oral cavity

Lateral tongue primordia

Medial tongue primordia

Mandibular
arch (I)

Hyoid
arch (II)

Arch
III

Arch
IV?

Multiple loci in the
placodal epithelium *of*
the arches gives rise to the
thyroid, parathyroid, and
thymus glands.

Laryngo-tracheal groove

0.25 mm

Section 85 brain *in situ*

Central neural structures labeled

DIENCEPHALON

THALAMUS/EPITHALAMUS

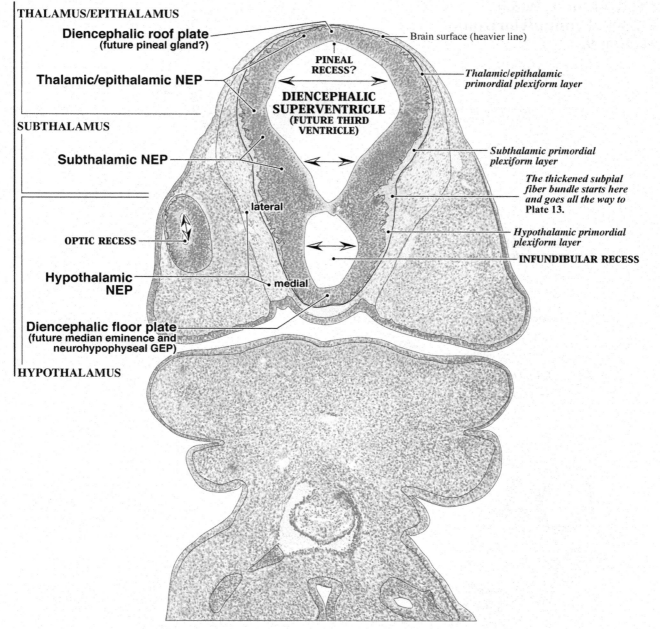

Diencephalic roof plate
(future pineal gland?)

Thalamic/epithalamic NEP

SUBTHALAMUS

Subthalamic NEP

lateral

OPTIC RECESS

Hypothalamic NEP

medial

Diencephalic floor plate
(future median eminence and
neurohypophyseal GEP)

HYPOTHALAMUS

— Brain surface (heavier line)

PINEAL RECESS?

DIENCEPHALIC SUPERVENTRICLE (FUTURE THIRD VENTRICLE)

Thalamic/epithalamic primordial plexiform layer

Subthalamic primordial plexiform layer

The thickened subpial fiber bundle starts here and goes all the way to Plate 13.

Hypothalamic primordial plexiform layer

INFUNDIBULAR RECESS

ABBREVIATIONS:
GEP - Glioepithelium
NEP - Neuroepithelium

Indicates expansion of the
NEP by *stockbuilding* and
final neurogenetic divisions.

FONT KEY:
VENTRICULAR DIVISIONS - CAPITALS
Germinal zone - Helvetica bold
Transient structure - Times bold italic
Permanent structure - Times Roman or **Bold**

PLATE 5A

CR 6.3 mm, GW 5.0
M2300, Frontal/Horizontal
Section 95

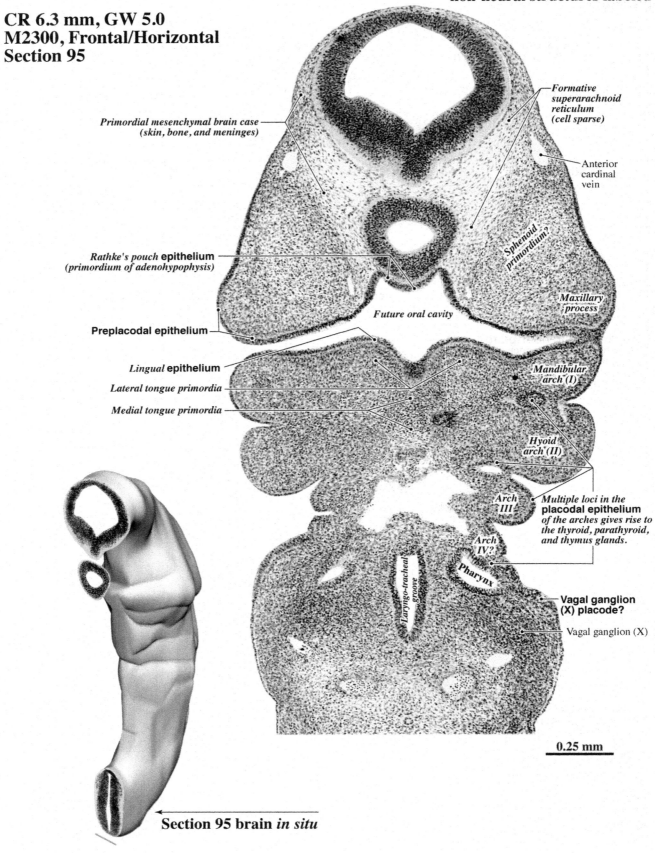

*Formative
superarachnoid
reticulum
(cell sparse)*

Primordial mesenchymal brain case
(skin, bone, and meninges)

Anterior
cardinal
vein

*Sphenoid
primordium?*

Rathke's pouch **epithelium**
(primordium of adenohypophysis)

*Maxillary
process*

Future oral cavity

Preplacodal epithelium

*Mandibular
arch (I)*

Lingual **epithelium**

Lateral tongue primordia

Medial tongue primordia

*Hyoid
arch (II)*

*Arch
III*

Multiple loci in the
placodal epithelium
*of the arches gives rise to
the thyroid, parathyroid,
and thymus glands.*

*Arch
IV?*

*Laryngo-tracheal
groove*

Pharynx

**Vagal ganglion
(X) placode?**

Vagal ganglion (X)

0.25 mm

Section 95 brain *in situ*

Central neural structures labeled

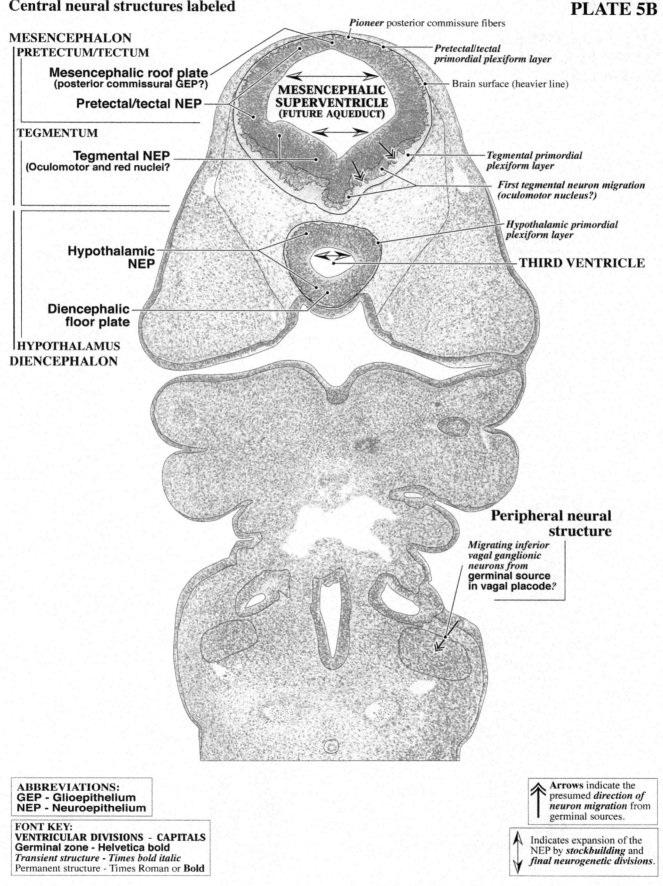

Pioneer posterior commissure fibers

Pretectal/tectal primordial plexiform layer

Brain surface (heavier line)

MESENCEPHALON
PRETECTUM/TECTUM

Mesencephalic roof plate
(posterior commissural GEP?)

Pretectal/tectal NEP

MESENCEPHALIC SUPERVENTRICLE (FUTURE AQUEDUCT)

TEGMENTUM

Tegmental NEP
(Oculomotor and red nuclei?

Tegmental primordial plexiform layer

First tegmental neuron migration (oculomotor nucleus?)

Hypothalamic primordial plexiform layer

Hypothalamic NEP

THIRD VENTRICLE

Diencephalic floor plate

HYPOTHALAMUS
DIENCEPHALON

Peripheral neural structure

Migrating inferior vagal ganglionic neurons from **germinal source in vagal placode?**

ABBREVIATIONS:
GEP - Glioepithelium
NEP - Neuroepithelium

FONT KEY:
VENTRICULAR DIVISIONS - CAPITALS
Germinal zone - Helvetica bold
Transient structure - Times bold italic
Permanent structure - Times Roman or **Bold**

Arrows indicate the presumed *direction of neuron migration* from germinal sources.

Indicates expansion of the NEP by *stockbuilding* and *final neurogenetic divisions*.

PLATE 6A

CR 6.3 mm, GW 5.0
M2300, Frontal/Horizontal
Section 115

**Peripheral neural
and non-neural
structures
labeled**

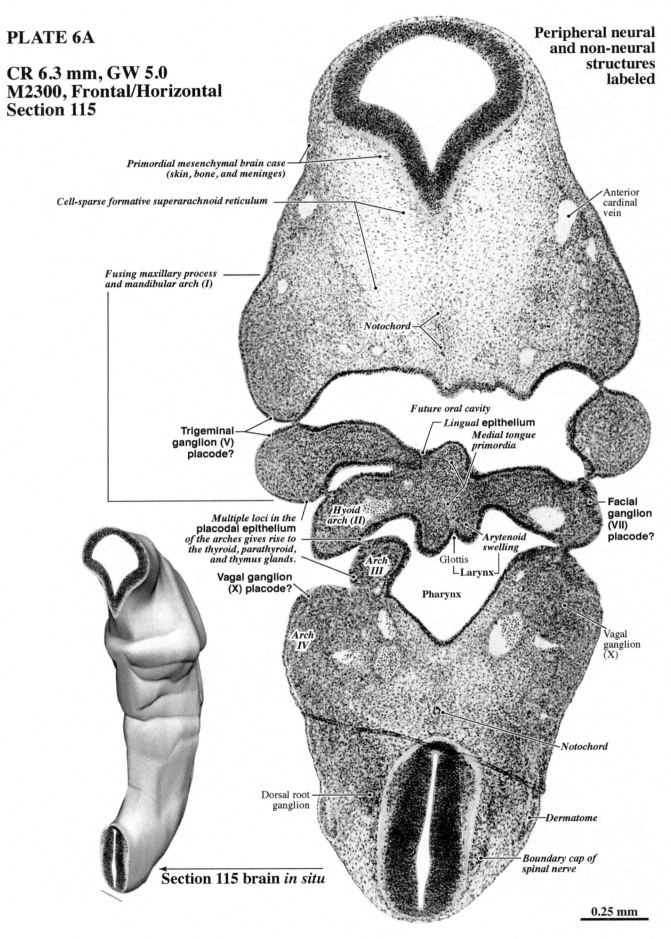

*Primordial mesenchymal brain case
(skin, bone, and meninges)*

Cell-sparse formative superarachnoid reticulum

Anterior
cardinal
vein

*Fusing maxillary process
and mandibular arch (I)*

Notochord

Future oral cavity
Lingual epithelium
*Medial tongue
primordia*

**Trigeminal
ganglion (V)
placode?**

**Facial
ganglion
(VII)
placode?**

*Hyoid
arch (II)*

Multiple loci in the
placodal epithelium
*of the arches gives rise to
the thyroid, parathyroid,
and thymus glands.*

*Arytenoid
swelling*

Glottis
└Larynx

**Vagal ganglion
(X) placode?**

*Arch
III*

Pharynx

Vagal
ganglion
(X)

*Arch
IV*

Dorsal root
ganglion

Notochord

Dermatome

Section 115 brain *in situ*

*Boundary cap of
spinal nerve*

0.25 mm

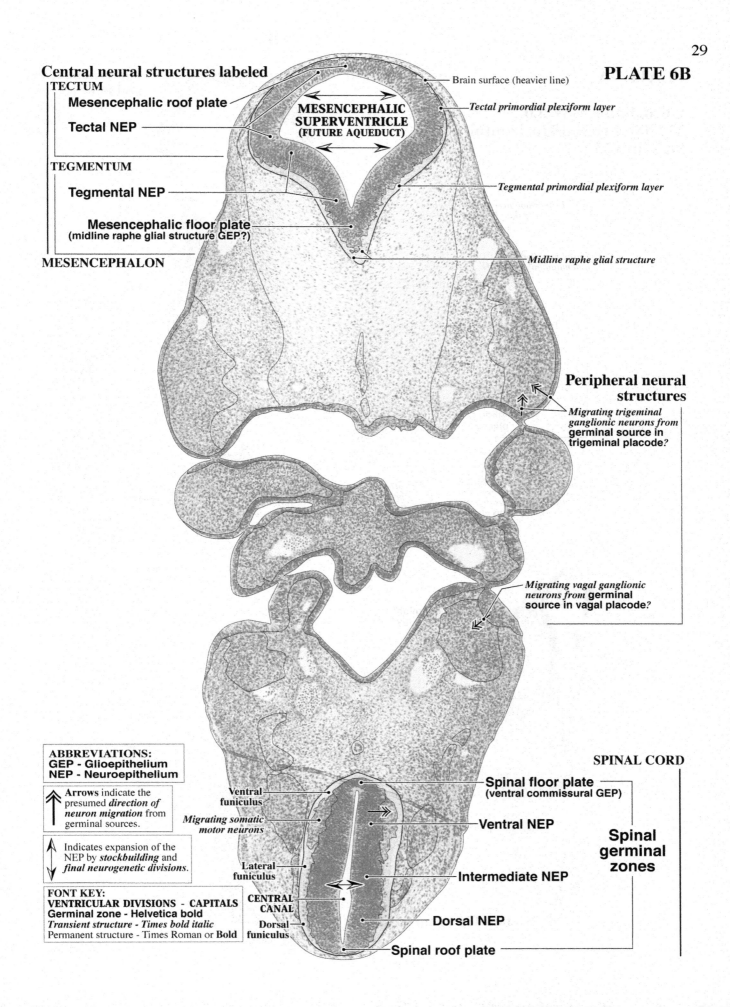

29

PLATE 6B

Central neural structures labeled

TECTUM
  Mesencephalic roof plate
  Tectal NEP

TEGMENTUM
  Tegmental NEP
  Mesencephalic floor plate
  (midline raphe glial structure GEP?)

MESENCEPHALON

Brain surface (heavier line)

MESENCEPHALIC SUPERVENTRICLE (FUTURE AQUEDUCT)

Tectal primordial plexiform layer

Tegmental primordial plexiform layer

Midline raphe glial structure

Peripheral neural structures

Migrating trigeminal ganglionic neurons from **germinal source in trigeminal placode**?

Migrating vagal ganglionic neurons from **germinal source in vagal placode**?

ABBREVIATIONS:
GEP - Glioepithelium
NEP - Neuroepithelium

Arrows indicate the presumed *direction of neuron migration* from germinal sources.

Indicates expansion of the NEP by *stockbuilding* and *final neurogenetic divisions*.

FONT KEY:
VENTRICULAR DIVISIONS - CAPITALS
Germinal zone - Helvetica bold
Transient structure - Times bold italic
Permanent structure - Times Roman or **Bold**

Ventral funiculus

Migrating somatic motor neurons

Lateral funiculus

CENTRAL CANAL

Dorsal funiculus

SPINAL CORD

Spinal floor plate
(ventral commissural GEP)

Ventral NEP

Intermediate NEP

Dorsal NEP

Spinal roof plate

Spinal germinal zones

PLATE 7A

CR 6.3 mm, GW 5.0
M2300, Frontal/Horizontal
Section 135

**Peripheral neural
and non-neural
structures
labeled**

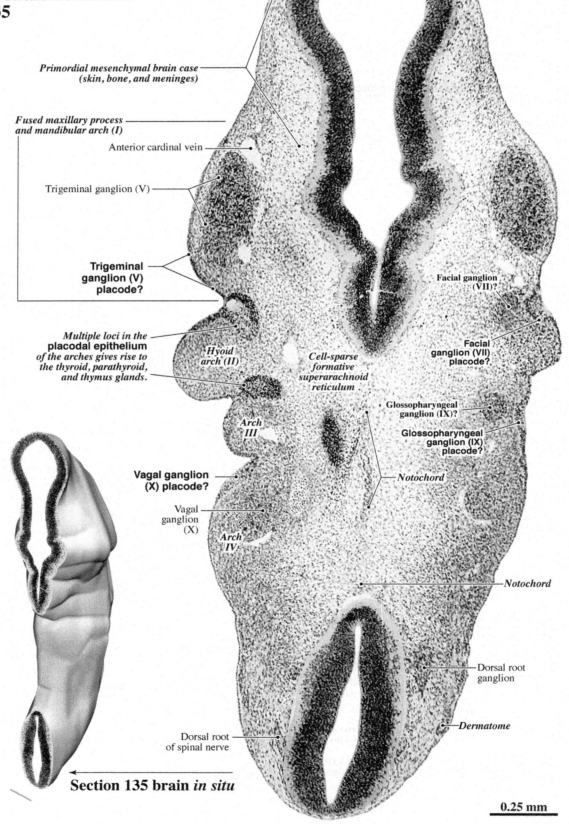

Primordial mesenchymal brain case
(skin, bone, and meninges)

Fused maxillary process
and mandibular arch (I)

Anterior cardinal vein

Trigeminal ganglion (V)

**Trigeminal
ganglion (V)
placode?**

Multiple loci in the
placodal epithelium
of the arches gives rise to
the thyroid, parathyroid,
and thymus glands.

Hyoid
arch (II)

Cell-sparse
formative
superarachnoid
reticulum

**Facial ganglion
(VII)?**

**Facial
ganglion (VII)
placode?**

**Glossopharyngeal
ganglion (IX)?**

**Glossopharyngeal
ganglion (IX)
placode?**

Arch
III

**Vagal ganglion
(X) placode?**

Vagal
ganglion
(X)

Arch
IV

Notochord

Notochord

Dorsal root
ganglion

Dorsal root
of spinal nerve

Dermatome

Section 135 brain *in situ*

0.25 mm

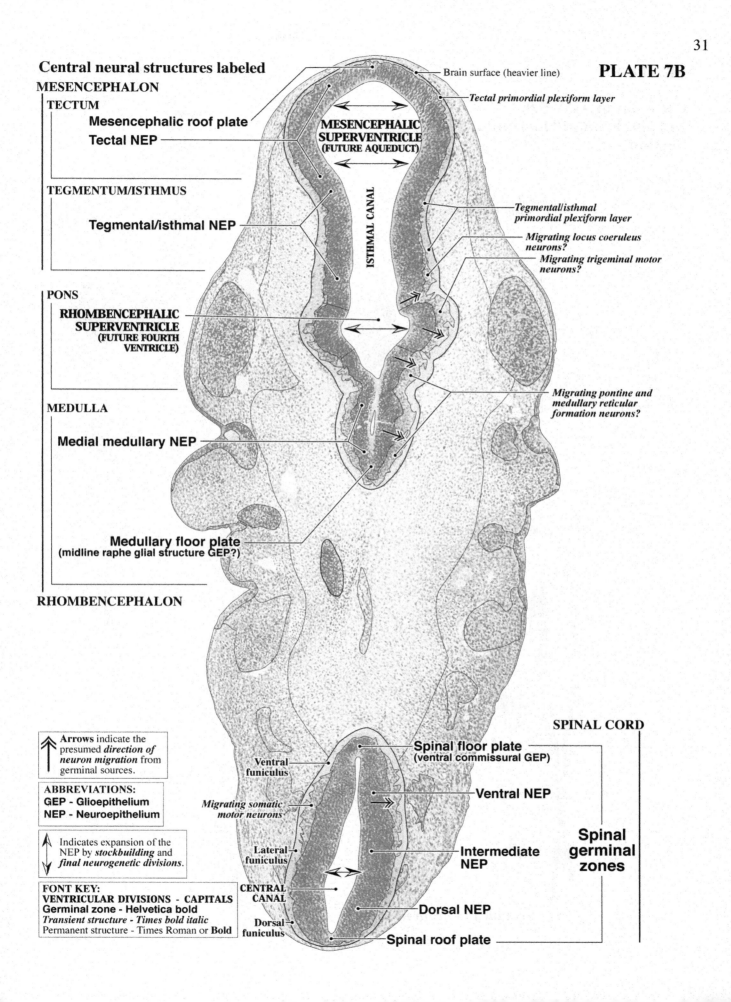

Central neural structures labeled

PLATE 7B

31

MESENCEPHALON

TECTUM

Mesencephalic roof plate

Tectal NEP

Brain surface (heavier line)

Tectal primordial plexiform layer

MESENCEPHALIC SUPERVENTRICLE (FUTURE AQUEDUCT)

ISTHMAL CANAL

TEGMENTUM/ISTHMUS

Tegmental/isthmal NEP

Tegmental/isthmal primordial plexiform layer

Migrating locus coeruleus neurons?

Migrating trigeminal motor neurons?

PONS

RHOMBENCEPHALIC SUPERVENTRICLE (FUTURE FOURTH VENTRICLE)

Migrating pontine and medullary reticular formation neurons?

MEDULLA

Medial medullary NEP

Medullary floor plate
(midline raphe glial structure GEP?)

RHOMBENCEPHALON

SPINAL CORD

Arrows indicate the presumed *direction of neuron migration* from germinal sources.

ABBREVIATIONS:
GEP - Glioepithelium
NEP - Neuroepithelium

Indicates expansion of the NEP by *stockbuilding* and *final neurogenetic divisions*.

FONT KEY:
VENTRICULAR DIVISIONS - CAPITALS
Germinal zone - Helvetica bold
Transient structure - Times bold italic
Permanent structure - Times Roman or **Bold**

Ventral funiculus

Migrating somatic motor neurons

Lateral funiculus

CENTRAL CANAL

Dorsal funiculus

Spinal floor plate
(ventral commissural GEP)

Ventral NEP

Intermediate NEP

Dorsal NEP

Spinal roof plate

Spinal germinal zones

32

PLATE 8A

CR 6.3 mm, GW 5.0
M2300, Frontal/Horizontal
Section 145

Peripheral neural
and non-neural
structures
labeled

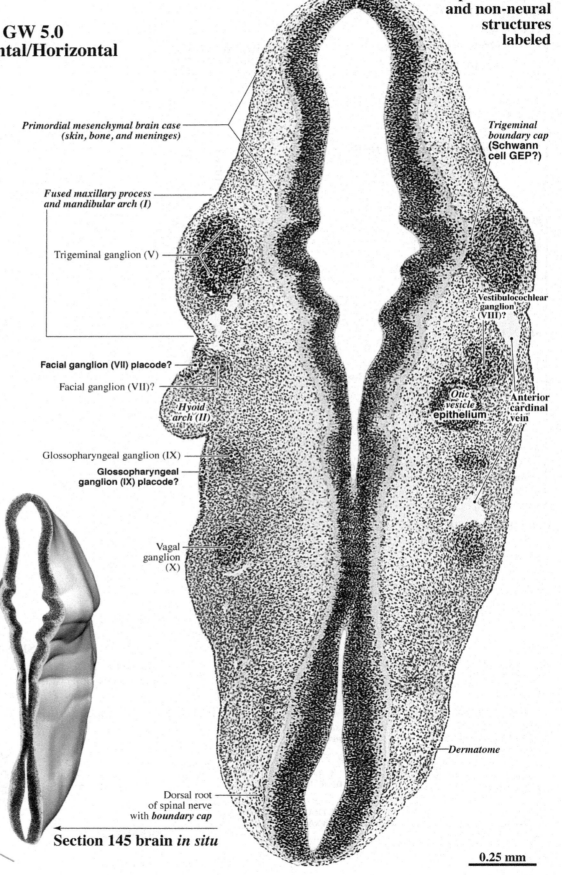

Primordial mesenchymal brain case
(skin, bone, and meninges)

Trigeminal
boundary cap
(Schwann
cell GEP?)

Fused maxillary process
and mandibular arch (I)

Trigeminal ganglion (V)

Vestibulocochlear
ganglion
(VIII)?

Facial ganglion (VII) placode?

Facial ganglion (VII)?

Otic
vesicle
epithelium

Anterior
cardinal
vein

Hyoid
arch (II)

Glossopharyngeal ganglion (IX)

Glossopharyngeal
ganglion (IX) placode?

Vagal
ganglion
(X)

Dermatome

Dorsal root
of spinal nerve
with *boundary cap*

Section 145 brain *in situ*

0.25 mm

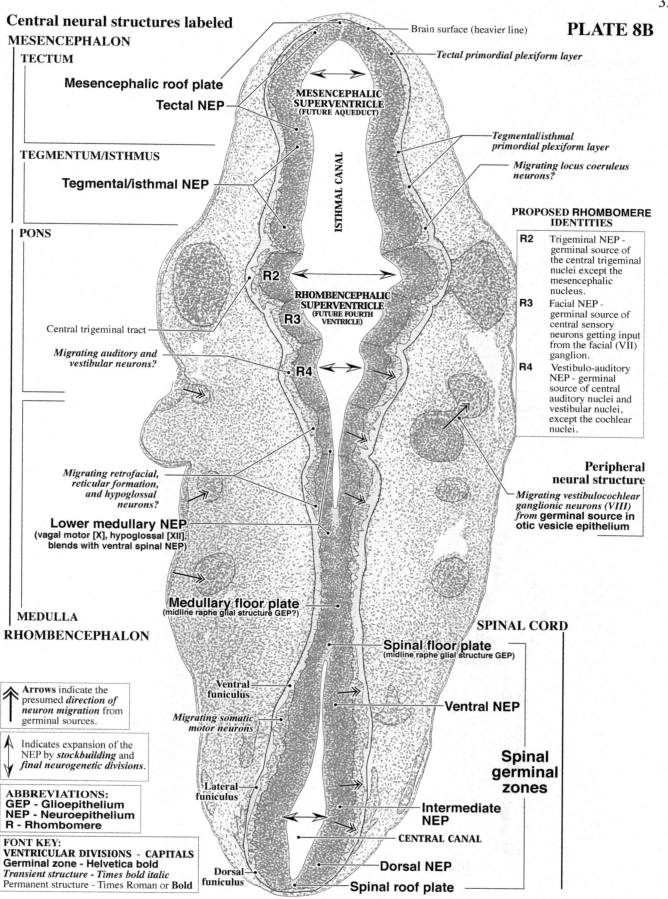

Central neural structures labeled

MESENCEPHALON

TECTUM

Mesencephalic roof plate

Tectal NEP

TEGMENTUM/ISTHMUS

Tegmental/isthmal NEP

PONS

Central trigeminal tract

Migrating auditory and vestibular neurons?

Migrating retrofacial, reticular formation, and hypoglossal neurons?

Lower medullary NEP
(vagal motor [X], hypoglossal [XII], blends with ventral spinal NEP)

MEDULLA

RHOMBENCEPHALON

Brain surface (heavier line)

Tectal primordial plexiform layer

MESENCEPHALIC SUPERVENTRICLE
(FUTURE AQUEDUCT)

ISTHMAL CANAL

Tegmental/isthmal primordial plexiform layer

Migrating locus coeruleus neurons?

R2

RHOMBENCEPHALIC SUPERVENTRICLE
(FUTURE FOURTH VENTRICLE)

R3

R4

PROPOSED RHOMBOMERE IDENTITIES

R2	Trigeminal NEP - germinal source of the central trigeminal nuclei except the mesencephalic nucleus.
R3	Facial NEP - germinal source of central sensory neurons getting input from the facial (VII) ganglion.
R4	Vestibulo-auditory NEP - germinal source of central auditory nuclei and vestibular nuclei, except the cochlear nuclei.

Peripheral neural structure

Migrating vestibulocochlear ganglionic neurons (VIII) from germinal source in otic vesicle epithelium

Medullary floor plate
(midline raphe glial structure GEP?)

SPINAL CORD

Spinal floor plate
(midline raphe glial structure GEP)

Ventral funiculus

Migrating somatic motor neurons

Lateral funiculus

Dorsal funiculus

Ventral NEP

Spinal germinal zones

Intermediate NEP

CENTRAL CANAL

Dorsal NEP

Spinal roof plate

Arrows indicate the presumed *direction of neuron migration* from germinal sources.

Indicates expansion of the NEP by *stockbuilding* and *final neurogenetic divisions*.

ABBREVIATIONS:
GEP - Glioepithelium
NEP - Neuroepithelium
R - Rhombomere

FONT KEY:
VENTRICULAR DIVISIONS - CAPITALS
Germinal zone - Helvetica bold
Transient structure - Times bold italic
Permanent structure - Times Roman or **Bold**

PLATE 8B

34

PLATE 9A

CR 6.3 mm
GW 5.0
M2300
Section 155

Peripheral neural
and non-neural
structures
labeled

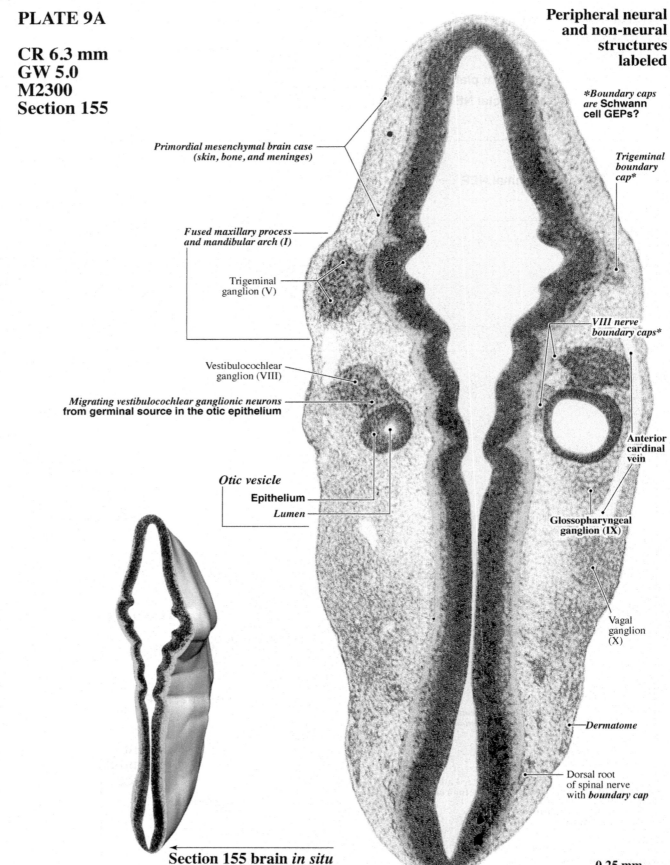

*Boundary caps
are Schwann
cell GEPs?

Primordial mesenchymal brain case
(skin, bone, and meninges)

Trigeminal
boundary
cap*

Fused maxillary process
and mandibular arch (I)

Trigeminal
ganglion (V)

VIII nerve
boundary caps*

Vestibulocochlear
ganglion (VIII)

Migrating vestibulocochlear ganglionic neurons
from germinal source in the otic epithelium

Anterior
cardinal
vein

Otic vesicle

Epithelium

Lumen

Glossopharyngeal
ganglion (IX)

Vagal
ganglion
(X)

Dermatome

Dorsal root
of spinal nerve
with boundary cap

Section 155 brain in situ

0.25 mm

PLATE 9B

Central neural structures labeled

MESENCEPHALON

TECTUM

Mesencephalic roof plate

Tectal NEP

ISTHMUS

Isthmal NEP

CEREBELLUM

Cerebellar NEP

PONS

Central trigeminal tract

R2

R3

R4

R5

R6

Migrating auditory and vestibular neurons

Migrating solitary nuclear neurons?

Lower medullary NEP
(vagal motor [X], hypoglossal [XII], blends with ventral spinal NEP)

MEDULLA

RHOMBENCEPHALON

Brain surface (heavier line)

Tectal primordial plexiform layer

MESENCEPHALIC SUPERVENTRICLE
(FUTURE AQUEDUCT)

Isthmal primordial plexiform layer

ISTHMAL CANAL

RHOMBENCEPHALIC SUPERVENTRICLE
(FUTURE FOURTH VENTRICLE)

Migrating retrofacial, reticular formation, and hypoglossal neurons?

Migrating somatic motor neurons

SPINAL CORD

Ventral gray

Ventral funiculus?

Ventral NEP

Lateral funiculus

Spinal germinal zones

Intermediate NEP

CENTRAL CANAL

Dorsal funiculus

Dorsal NEP

Spinal roof plate

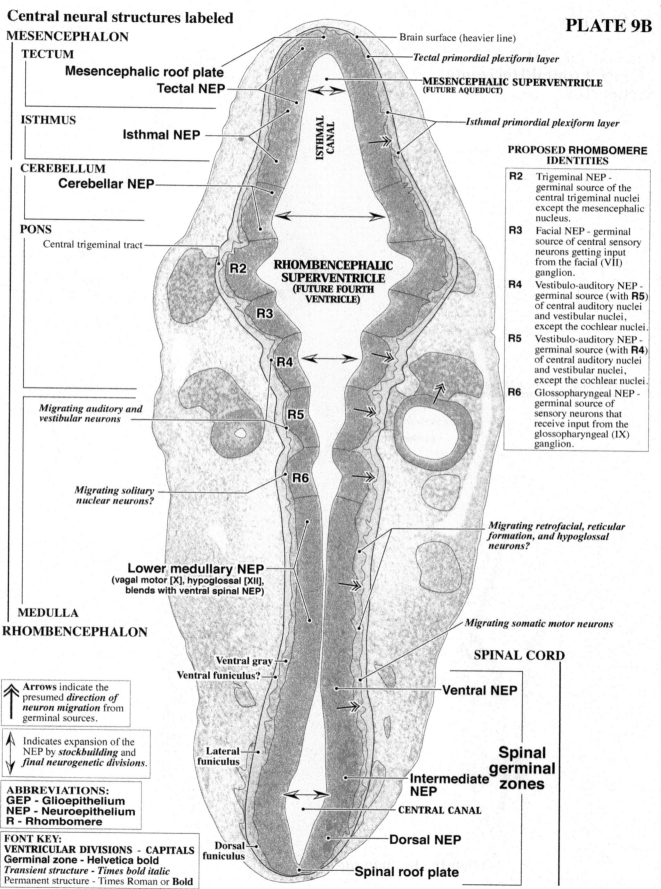

PROPOSED RHOMBOMERE IDENTITIES

R2 Trigeminal NEP - germinal source of the central trigeminal nuclei except the mesencephalic nucleus.

R3 Facial NEP - germinal source of central sensory neurons getting input from the facial (VII) ganglion.

R4 Vestibulo-auditory NEP - germinal source (with **R5**) of central auditory nuclei and vestibular nuclei, except the cochlear nuclei.

R5 Vestibulo-auditory NEP - germinal source (with **R4**) of central auditory nuclei and vestibular nuclei, except the cochlear nuclei.

R6 Glossopharyngeal NEP - germinal source of sensory neurons that receive input from the glossopharyngeal (IX) ganglion.

Arrows indicate the presumed *direction of neuron migration* from germinal sources.

Indicates expansion of the NEP by *stockbuilding* and *final neurogenetic divisions*.

ABBREVIATIONS:
GEP - Glioepithelium
NEP - Neuroepithelium
R - Rhombomere

FONT KEY:
VENTRICULAR DIVISIONS - CAPITALS
Germinal zone - Helvetica bold
Transient structure - Times bold italic
Permanent structure - Times Roman or **Bold**

36

PLATE 10A

CR 6.3 mm, GW 5.0
M2300, Frontal/Horizontal
Section 165

Peripheral neural and non-neural structures labeled

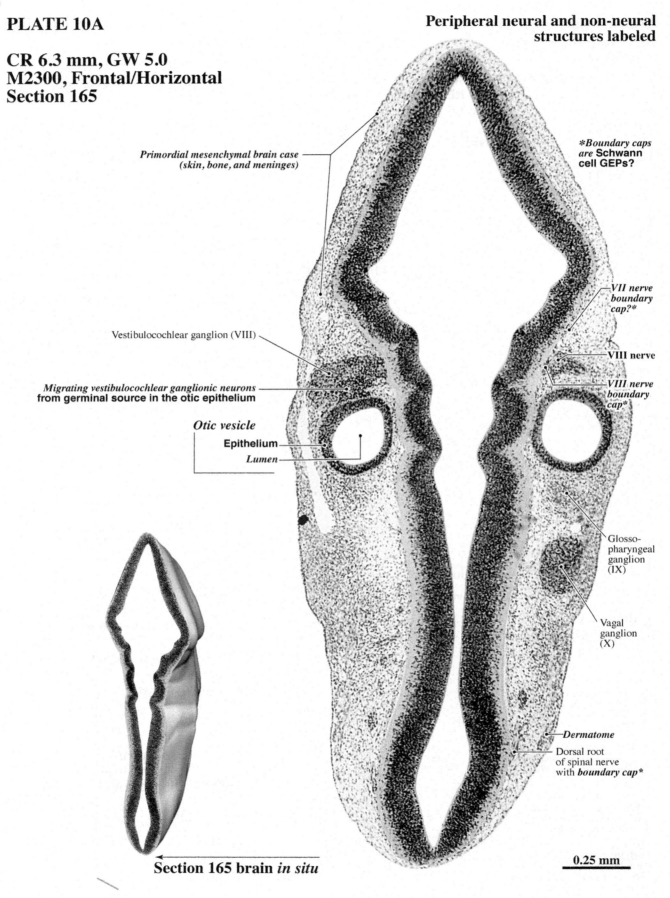

Primordial mesenchymal brain case
(skin, bone, and meninges)

*Boundary caps
are Schwann
cell GEPs?*

VII nerve
boundary
cap?*

VIII nerve

VIII nerve
boundary
cap*

Vestibulocochlear ganglion (VIII)

Migrating vestibulocochlear ganglionic neurons
from germinal source in the otic epithelium

Otic vesicle
Epithelium
Lumen

Glosso-
pharyngeal
ganglion
(IX)

Vagal
ganglion
(X)

Dermatome
Dorsal root
of spinal nerve
with *boundary cap**

Section 165 brain in situ

0.25 mm

Central neural structures labeled

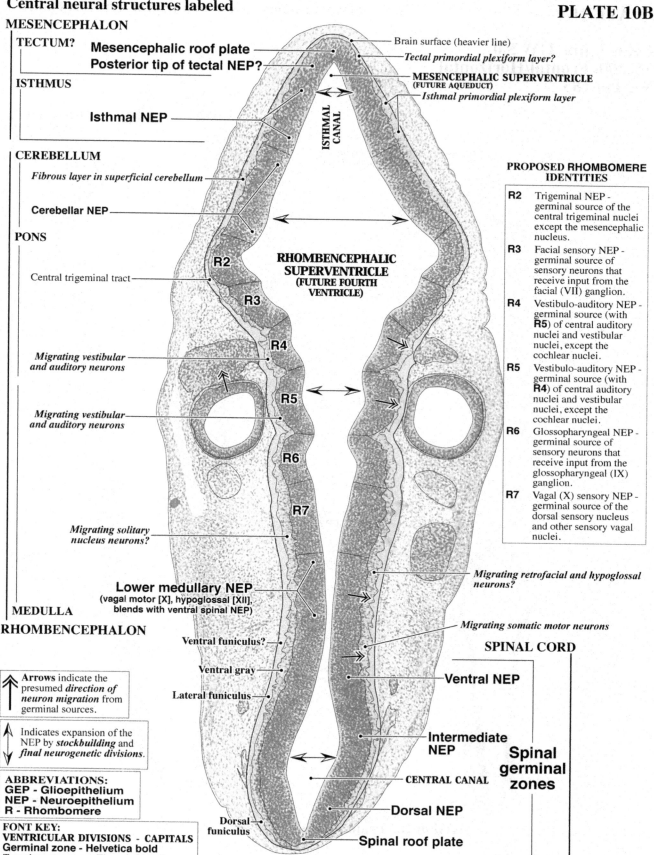

MESENCEPHALON

TECTUM?
- Mesencephalic roof plate
- Posterior tip of tectal NEP?

ISTHMUS
- Isthmal NEP

CEREBELLUM
- *Fibrous layer in superficial cerebellum*
- Cerebellar NEP

PONS
- Central trigeminal tract
- *Migrating vestibular and auditory neurons*
- *Migrating vestibular and auditory neurons*

MEDULLA

RHOMBENCEPHALON
- *Migrating solitary nucleus neurons?*
- Lower medullary NEP (vagal motor [X], hypoglossal [XII], blends with ventral spinal NEP)
- Ventral funiculus?
- Ventral gray
- Lateral funiculus

Brain surface (heavier line)
Tectal primordial plexiform layer?

MESENCEPHALIC SUPERVENTRICLE (FUTURE AQUEDUCT)
Isthmal primordial plexiform layer

ISTHMAL CANAL

RHOMBENCEPHALIC SUPERVENTRICLE (FUTURE FOURTH VENTRICLE)

R2
R3
R4
R5
R6
R7

Migrating retrofacial and hypoglossal neurons?

Migrating somatic motor neurons

SPINAL CORD

Ventral NEP

Intermediate NEP

Spinal germinal zones

CENTRAL CANAL

Dorsal NEP

Spinal roof plate

Dorsal funiculus

PROPOSED RHOMBOMERE IDENTITIES

R2 Trigeminal NEP - germinal source of the central trigeminal nuclei except the mesencephalic nucleus.

R3 Facial sensory NEP - germinal source of sensory neurons that receive input from the facial (VII) ganglion.

R4 Vestibulo-auditory NEP - germinal source (with **R5**) of central auditory nuclei and vestibular nuclei, except the cochlear nuclei.

R5 Vestibulo-auditory NEP - germinal source (with **R4**) of central auditory nuclei and vestibular nuclei, except the cochlear nuclei.

R6 Glossopharyngeal NEP - germinal source of sensory neurons that receive input from the glossopharyngeal (IX) ganglion.

R7 Vagal (X) sensory NEP - germinal source of the dorsal sensory nucleus and other sensory vagal nuclei.

Arrows indicate the presumed *direction of neuron migration* from germinal sources.

Indicates expansion of the NEP by *stockbuilding* and *final neurogenetic divisions*.

ABBREVIATIONS:
GEP - Glioepithelium
NEP - Neuroepithelium
R - Rhombomere

FONT KEY:
VENTRICULAR DIVISIONS - CAPITALS
Germinal zone - Helvetica bold
Transient structure - Times bold italic
Permanent structure - Times Roman or **Bold**

38

Peripheral neural and non-neural structures labeled

CR 6.3 mm, GW 5.0
M2300, Frontal/Horizontal
Section 185

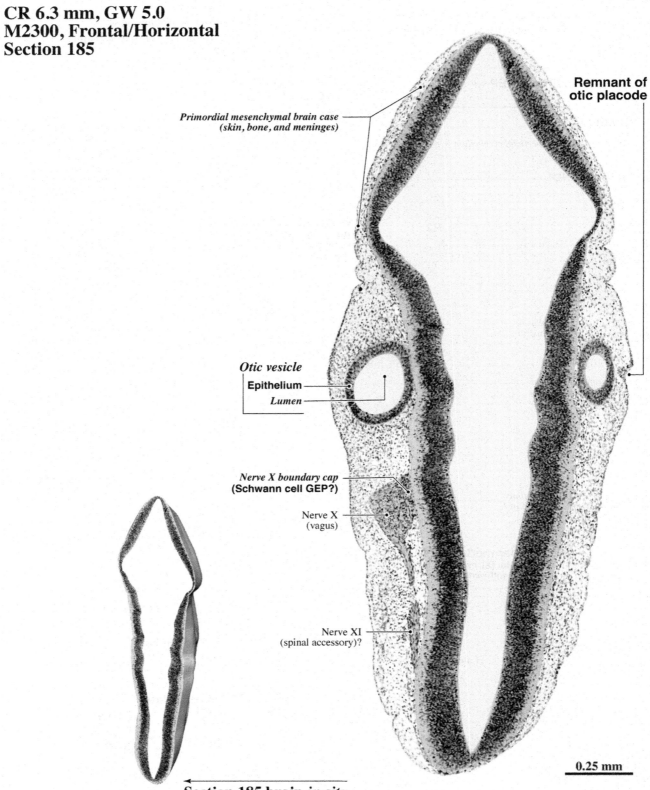

Remnant of
otic placode

Primordial mesenchymal brain case
(skin, bone, and meninges)

Otic vesicle

Epithelium

Lumen

Nerve X boundary cap
(Schwann cell GEP?)

Nerve X
(vagus)

Nerve XI
(spinal accessory)?

0.25 mm

Section 185 brain *in situ*

Central neural structures labeled

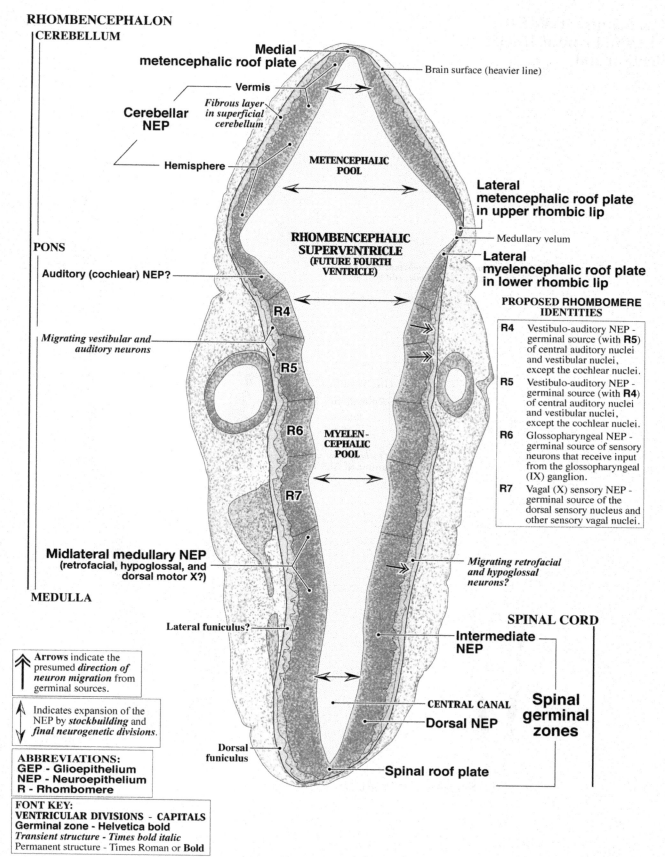

RHOMBENCEPHALON

CEREBELLUM

Medial
metencephalic roof plate

Brain surface (heavier line)

Vermis

Fibrous layer in superficial cerebellum

Cerebellar NEP

Hemisphere

METENCEPHALIC POOL

Lateral metencephalic roof plate in upper rhombic lip

Medullary velum

RHOMBENCEPHALIC SUPERVENTRICLE (FUTURE FOURTH VENTRICLE)

Lateral myelencephalic roof plate in lower rhombic lip

PONS

Auditory (cochlear) NEP?

Migrating vestibular and auditory neurons

R4

R5

R6

MYELEN-CEPHALIC POOL

R7

PROPOSED RHOMBOMERE IDENTITIES

R4	Vestibulo-auditory NEP - germinal source (with **R5**) of central auditory nuclei and vestibular nuclei, except the cochlear nuclei.
R5	Vestibulo-auditory NEP - germinal source (with **R4**) of central auditory nuclei and vestibular nuclei, except the cochlear nuclei.
R6	Glossopharyngeal NEP - germinal source of sensory neurons that receive input from the glossopharyngeal (IX) ganglion.
R7	Vagal (X) sensory NEP - germinal source of the dorsal sensory nucleus and other sensory vagal nuclei.

Midlateral medullary NEP
(retrofacial, hypoglossal, and dorsal motor X?)

Migrating retrofacial and hypoglossal neurons?

MEDULLA

SPINAL CORD

Lateral funiculus?

Intermediate NEP

Arrows indicate the presumed *direction of neuron migration* from germinal sources.

Indicates expansion of the NEP by *stockbuilding* and *final neurogenetic divisions*.

CENTRAL CANAL

Dorsal NEP

Spinal germinal zones

ABBREVIATIONS:
GEP - Glioepithelium
NEP - Neuroepithelium
R - Rhombomere

Dorsal funiculus

Spinal roof plate

FONT KEY:
VENTRICULAR DIVISIONS - CAPITALS
Germinal zone - Helvetica bold
Transient structure - Times bold italic
Permanent structure - Times Roman or **Bold**

40

PLATE 12A

CR 6.3 mm, GW 5.0
M2300, Frontal/Horizontal
Section 200

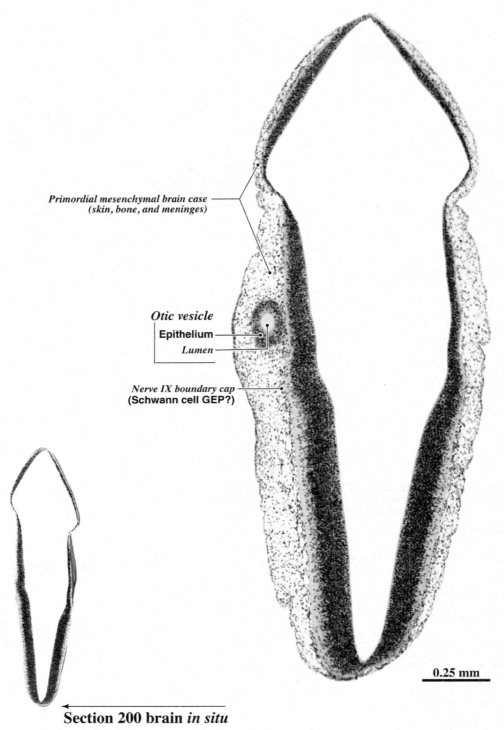

Primordial mesenchymal brain case
(skin, bone, and meninges)

Otic vesicle

Epithelium

Lumen

Nerve IX boundary cap
(Schwann cell GEP?)

0.25 mm

Section 200 brain *in situ*

Central neural structures labeled

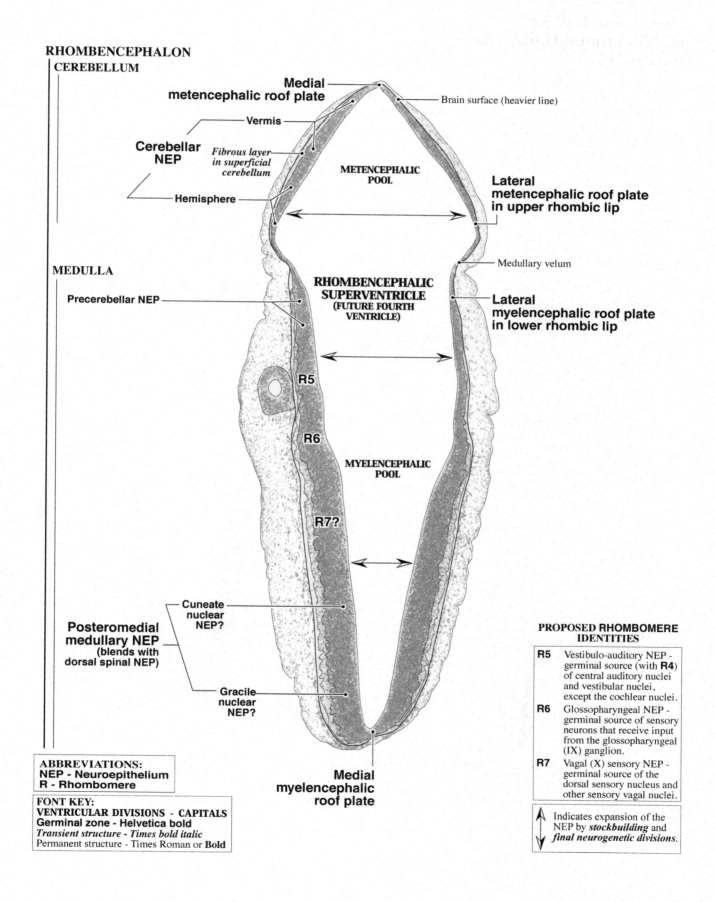

RHOMBENCEPHALON
CEREBELLUM

Medial metencephalic roof plate

Brain surface (heavier line)

Vermis

Cerebellar NEP

Fibrous layer in superficial cerebellum

METENCEPHALIC POOL

Hemisphere

Lateral metencephalic roof plate in upper rhombic lip

MEDULLA

Medullary velum

Precerebellar NEP

RHOMBENCEPHALIC SUPERVENTRICLE (FUTURE FOURTH VENTRICLE)

Lateral myelencephalic roof plate in lower rhombic lip

R5

R6

MYELENCEPHALIC POOL

R7?

Cuneate nuclear NEP?

Posteromedial medullary NEP (blends with dorsal spinal NEP)

Gracile nuclear NEP?

Medial myelencephalic roof plate

ABBREVIATIONS:
NEP - Neuroepithelium
R - Rhombomere

FONT KEY:
VENTRICULAR DIVISIONS - CAPITALS
Germinal zone - Helvetica bold
Transient structure - Times bold italic
Permanent structure - Times Roman or **Bold**

PROPOSED RHOMBOMERE IDENTITIES

R5	Vestibulo-auditory NEP - germinal source (with **R4**) of central auditory nuclei and vestibular nuclei, except the cochlear nuclei.
R6	Glossopharyngeal NEP - germinal source of sensory neurons that receive input from the glossopharyngeal (IX) ganglion.
R7	Vagal (X) sensory NEP - germinal source of the dorsal sensory nucleus and other sensory vagal nuclei.

Indicates expansion of the NEP by *stockbuilding* and *final neurogenetic divisions*.

PLATE 13A

CR 6.3 mm, GW 5.0
M2300, Frontal/Horizontal
Section 210

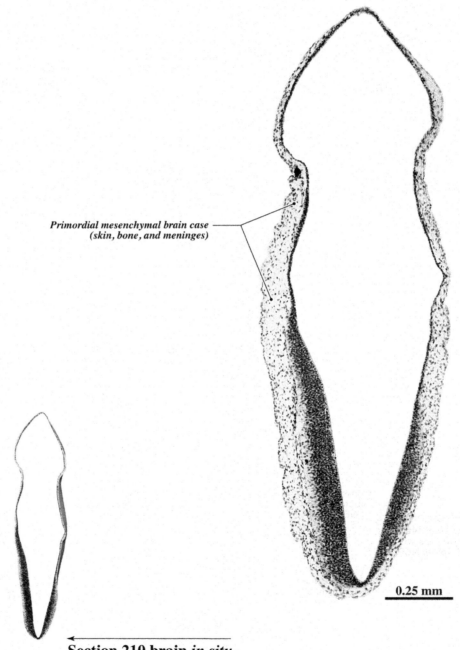

Primordial mesenchymal brain case
(skin, bone, and meninges)

0.25 mm

Section 210 brain *in situ*

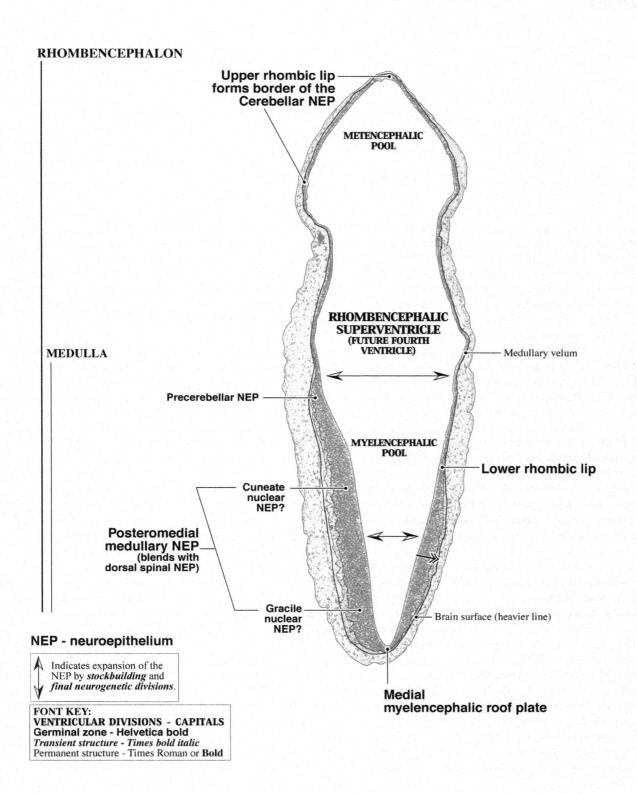

RHOMBENCEPHALON

Upper rhombic lip forms border of the Cerebellar NEP

METENCEPHALIC POOL

MEDULLA

RHOMBENCEPHALIC SUPERVENTRICLE (FUTURE FOURTH VENTRICLE)

Medullary velum

Precerebellar NEP

MYELENCEPHALIC POOL

Lower rhombic lip

Cuneate nuclear NEP?

Posteromedial medullary NEP (blends with dorsal spinal NEP)

Gracile nuclear NEP?

Brain surface (heavier line)

Medial myelencephalic roof plate

NEP - neuroepithelium

Indicates expansion of the NEP by *stockbuilding* and *final neurogenetic divisions*.

FONT KEY:
VENTRICULAR DIVISIONS - CAPITALS
Germinal zone - Helvetica bold
Transient structure - Times bold italic
Permanent structure - Times Roman or **Bold**

PART III: C8966
CR 7.1 mm (GW 5.5)
Sagittal

Carnegie collection specimen #8966 (designated here as C8996) with a 7.1-mm crown-rump length (CR) is estimated to be at gestation week (GW) 5.5. C8966 was preserved in Zenker's fixative, embedded in a celloidin/paraffin mix, and was cut in 10-μm sagittal sections that were stained with hematoxylin and eosin. Various orientations of the computer-aided 3-D reconstructions of C8314's brain are used to show the gross external features of a GW5.5 brain (**Fig. 11**). Like most sagittally-cut specimens, C8966's sections are not parallel to the midline: **Figure 11** shows the approximate rotations in front (**B**) and back views (**C**). We photographed 29 sections at low magnification from the left to right sides of the brain. Seven of the sections, mainly from the left side, are illustrated in Plates **14AB** to **20AB**. Each illustrated section shows the brain with all surrounding tissues. Labels in **A Plates** (normal-contrast images) identify the approximate midline, non-neural structures, peripheral neural structures and brain ventricular divisions; labels in **B Plates** (low-contrast images) identify neural structures. **Plates 21-22AB** show hign-magnification views of the rhombencephalon.

The anterior part of the prosencephalon can be tentatively identified as the future telencephalon as an enlargement in front of the diencephalon. Most parts of the prosencephalon NEP are stockbuilding their stem cells, but some neurogenesis is taking place in the hypothalamus and subthalamus. There are a few pioneering postmitotic neurons outside the NEP of the basal telencephalon. The sagittal plane is best to show the thin lamina terminalis (marking the last closure of the anterior neuropore) between the preoptic area and the optic evagination. The optic evagination is now better defined at its terminus with a thicker retinal invagination and a thinner pigment epithelium starting to form an eyeball. The NEP of the thalamus is still stockbuilding, but the epithalamic NEP is entering the neurogenic stage.

The mesencephalic flexure area has a thick tegmental floor with accumulations of postmitotic neurons, probably interacting with subpial axons growing in from the sensory cranial nerves (mainly V and VII) and centrally originating axons from the spinal cord and medulla. The tectal NEPs are stockbuilding their pool of progenitors and few to no neurons are being generated there.

The big story in the rhombencephalon is that rhombomere NEPs are now well into their neurogenetic stages. Every rhombomere has postmitotic neurons migrating out, and the medullary and pontine floors are thickening. There is fascinating interaction with the ganglia in the periphery. The large trigeminal ganglion is sending fibers into the brain outside R2. Facial and vestibulocochlear nerve fibers are invading the brain outside R3 and R4. These nerves have boundary caps and presumptive Schwann glia are moving among the ingrowing fibers. Once the nerve fibers are inside the brain, there are few interspersed cells that may be considered glia. The glossopharyngeal ganglion is small but definite, and both superior and inferior parts of the vagal ganglion can be identified. The trigeminal, facial, glossopharyngeal and vagal ganglia can be located either attached to or just outside placodes in the pharyngeal arches (branchial placodes). The large otic vesicle is intimately associated with the enlarging vestibulocochlear ganglion and is its presumed germinal source.

Things are beginning to happen in the cerebellum. There is a laminated transitional field of alternating fibers and cells outside the very thick cerebellar NEP. The cells are most likely the early-generated deep nuclear neurons in the fastigial and interpositus deep nuclei. There is still lots of stockbuilding going on in the cerebellar NEP and the exit of a few postmitotic deep neurons does not seem to deplete the growth of the cerebellar NEP as it builds up the enormous numbers of Purkinje cell progenitors and progenitors of the massive dentate deep nucleus.

EXTERNAL FEATURES OF THE GW5.5-6.0 BRAIN

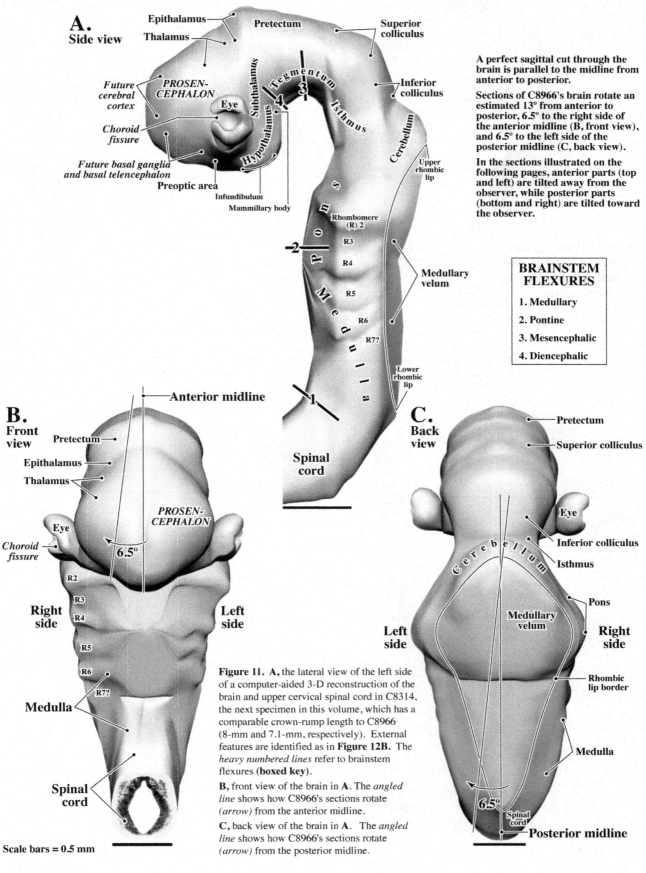

A. Side view

Epithalamus
Thalamus
Pretectum
Superior colliculus
Future cerebral cortex
PROSEN-CEPHALON
Subthalamus
Tegmentum
Isthmus
Inferior colliculus
Choroid fissure
Eye
Hypothalamus
Cerebellum
Future basal ganglia and basal telencephalon
Preoptic area
Infundibulum
Mammillary body
Upper rhombic lip
Pons
Rhombomere (R) 2
R3
R4
R5
R6
R7?
Medullary velum
Medulla
Lower rhombic lip
Spinal cord

A perfect sagittal cut through the brain is parallel to the midline from anterior to posterior.

Sections of C8966's brain rotate an estimated 13° from anterior to posterior, 6.5° to the right side of the anterior midline (B, front view), and 6.5° to the left side of the posterior midline (C, back view).

In the sections illustrated on the following pages, anterior parts (top and left) are tilted away from the observer, while posterior parts (bottom and right) are tilted toward the observer.

BRAINSTEM FLEXURES
1. Medullary
2. Pontine
3. Mesencephalic
4. Diencephalic

B. Front view

Anterior midline
Pretectum
Epithalamus
Thalamus
Eye
Choroid fissure
PROSEN-CEPHALON
6.5°
Right side
Left side
R2
R3
R4
R5
R6
R7?
Medulla
Spinal cord

Scale bars = 0.5 mm

Figure 11. A, the lateral view of the left side of a computer-aided 3-D reconstruction of the brain and upper cervical spinal cord in C8314, the next specimen in this volume, which has a comparable crown-rump length to C8966 (8-mm and 7.1-mm, respectively). External features are identified as in **Figure 12B.** The *heavy numbered lines* refer to brainstem flexures (**boxed key**).

B, front view of the brain in **A**. The *angled line* shows how C8966's sections rotate *(arrow)* from the anterior midline.

C, back view of the brain in **A**. The *angled line* shows how C8966's sections rotate *(arrow)* from the posterior midline.

C. Back view

Pretectum
Superior colliculus
Eye
Inferior colliculus
Isthmus
Cerebellum
Pons
Left side
Medullary velum
Right side
Rhombic lip border
Medulla
6.5°
Spinal cord
Posterior midline

PLATE 14A

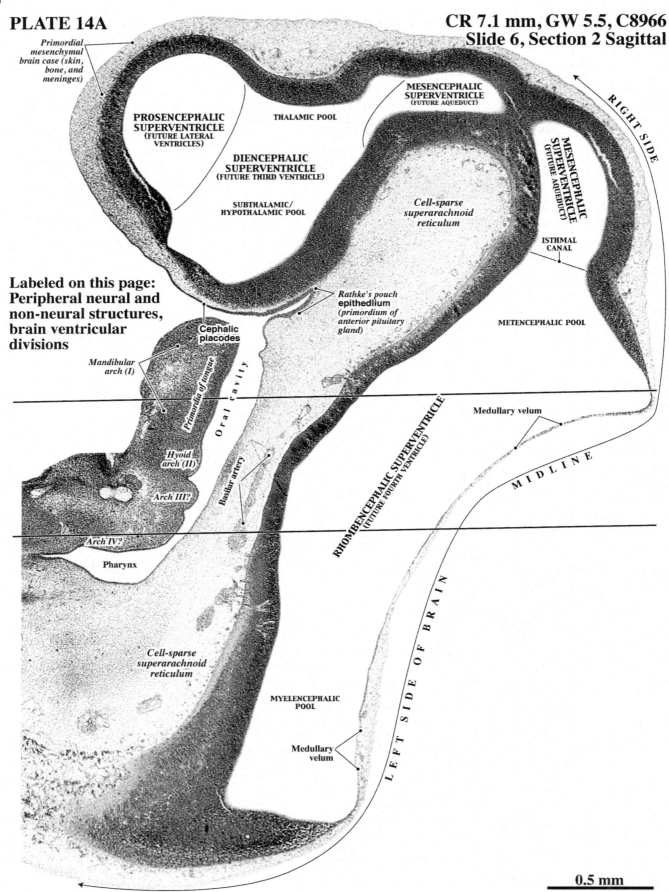

*Primordial
mesenchymal
brain case (skin,
bone, and
meninges)*

MESENCEPHALIC
SUPERVENTRICLE
(FUTURE AQUEDUCT)

THALAMIC POOL

RIGHT SIDE

PROSENCEPHALIC
SUPERVENTRICLE
(FUTURE LATERAL
VENTRICLES)

DIENCEPHALIC
SUPERVENTRICLE
(FUTURE THIRD VENTRICLE)

MESENCEPHALIC
SUPERVENTRICLE
(FUTURE AQUEDUCT)

SUBTHALAMIC/
HYPOTHALAMIC POOL

*Cell-sparse
superarachnoid
reticulum*

ISTHMAL
CANAL

**Labeled on this page:
Peripheral neural and
non-neural structures,
brain ventricular
divisions**

Cephalic
placodes

Rathke's pouch
epithedlium
*(primordium of
anterior pituitary
gland)*

METENCEPHALIC POOL

*Mandibular
arch (I)*

Oral cavity

Primordia of tongue

Medullary velum

*Hyoid
arch (II)*

Basilar artery

MIDLINE

Arch III?

RHOMBENCEPHALIC SUPERVENTRICLE
(FUTURE FOURTH VENTRICLE)

Arch IV?

Pharynx

*Cell-sparse
superarachnoid
reticulum*

LEFT SIDE OF BRAIN

MYELENCEPHALIC
POOL

Medullary
velum

0.5 mm

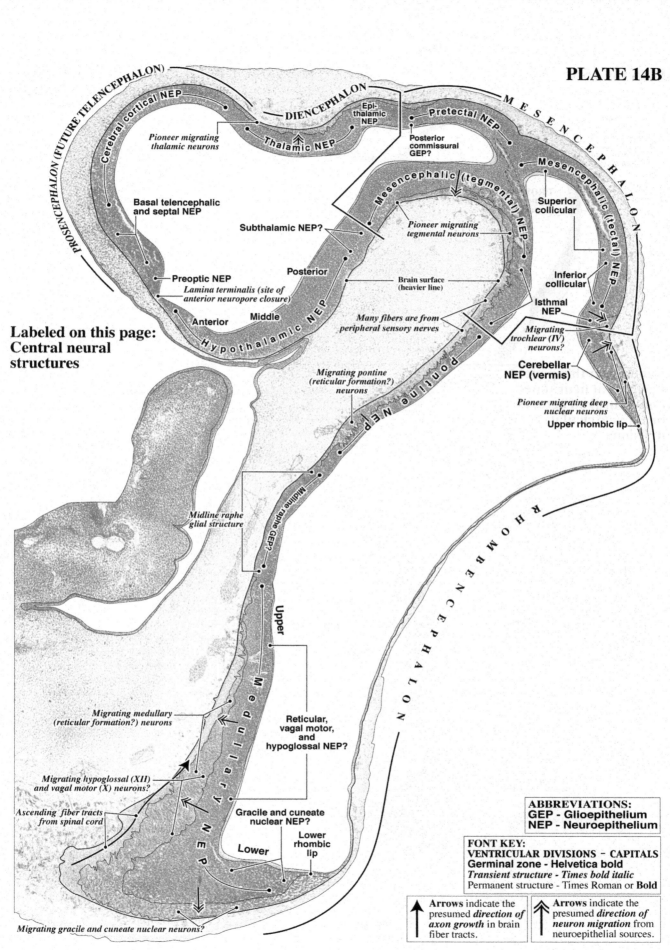

PROSENCEPHALON (FUTURE TELENCEPHALON)

Cerebral cortical NEP

Pioneer migrating thalamic neurons

DIENCEPHALON

Thalamic NEP

Epi-thalamic NEP

Pretectal NEP

Posterior commissural GEP?

MESENCEPHALON

Mesencephalic (tegmental) NEP

Mesencephalic (tectal) NEP

Superior collicular

Pioneer migrating tegmental neurons

Inferior collicular

Isthmal NEP

Basal telencephalic and septal NEP

Subthalamic NEP?

Posterior

Preoptic NEP

Lamina terminalis (site of anterior neuropore closure)

Anterior

Middle

Hypothalamic NEP

Brain surface (heavier line)

Many fibers are from peripheral sensory nerves

Migrating trochlear (IV) neurons?

Cerebellar NEP (vermis)

Pioneer migrating deep nuclear neurons

Upper rhombic lip

Labeled on this page: Central neural structures

Migrating pontine (reticular formation?) neurons

Pontine NEP

Midline raphe glial structure

Midline raphe GEP?

RHOMBENCEPHALON

Upper

Medullary NEP

Migrating medullary (reticular formation?) neurons

Reticular, vagal motor, and hypoglossal NEP?

Migrating hypoglossal (XII) and vagal motor (X) neurons?

Ascending fiber tracts from spinal cord

Gracile and cuneate nuclear NEP?

Lower

Lower rhombic lip

Migrating gracile and cuneate nuclear neurons?

ABBREVIATIONS:
GEP - Glioepithelium
NEP - Neuroepithelium

FONT KEY:
VENTRICULAR DIVISIONS – CAPITALS
Germinal zone - Helvetica bold
Transient structure - Times bold italic
Permanent structure - Times Roman or **Bold**

Arrows indicate the presumed ***direction of axon growth*** in brain fiber tracts.

Arrows indicate the presumed ***direction of neuron migration*** from neuroepithelial sources.

PLATE 15A

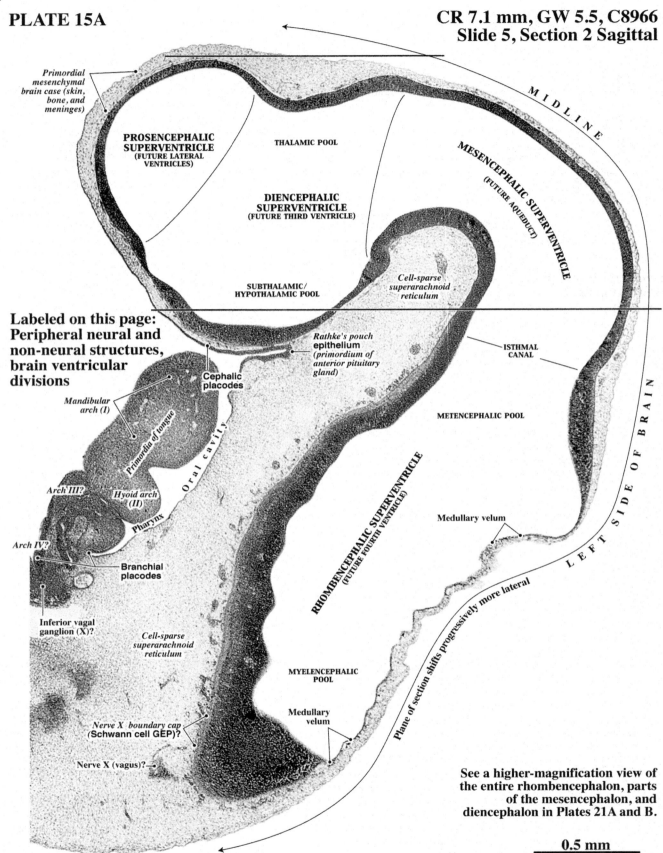

Primordial mesenchymal brain case (skin, bone, and meninges)

PROSENCEPHALIC SUPERVENTRICLE
(FUTURE LATERAL VENTRICLES)

THALAMIC POOL

M I D L I N E

MESENCEPHALIC SUPERVENTRICLE
(FUTURE AQUEDUCT)

DIENCEPHALIC SUPERVENTRICLE
(FUTURE THIRD VENTRICLE)

SUBTHALAMIC / HYPOTHALAMIC POOL

Cell-sparse superarachnoid reticulum

Labeled on this page: Peripheral neural and non-neural structures, brain ventricular divisions

Rathke's pouch epithelium *(primordium of anterior pituitary gland)*

ISTHMAL CANAL

Cephalic placodes

Mandibular arch (I)

Primordia of tongue

METENCEPHALIC POOL

L E F T   S I D E   O F   B R A I N

Arch III?

Hyoid arch (II)

Oral cavity

Pharynx

RHOMBENCEPHALIC SUPERVENTRICLE
(FUTURE FOURTH VENTRICLE)

Medullary velum

Arch IV?

Branchial placodes

Inferior vagal ganglion (X)?

Cell-sparse superarachnoid reticulum

MYELENCEPHALIC POOL

Plane of section shifts progressively more lateral

Medullary velum

Nerve X boundary cap (Schwann cell GEP)?

Nerve X (vagus)?

See a higher-magnification view of the entire rhombencephalon, parts of the mesencephalon, and diencephalon in Plates 21A and B.

0.5 mm

49

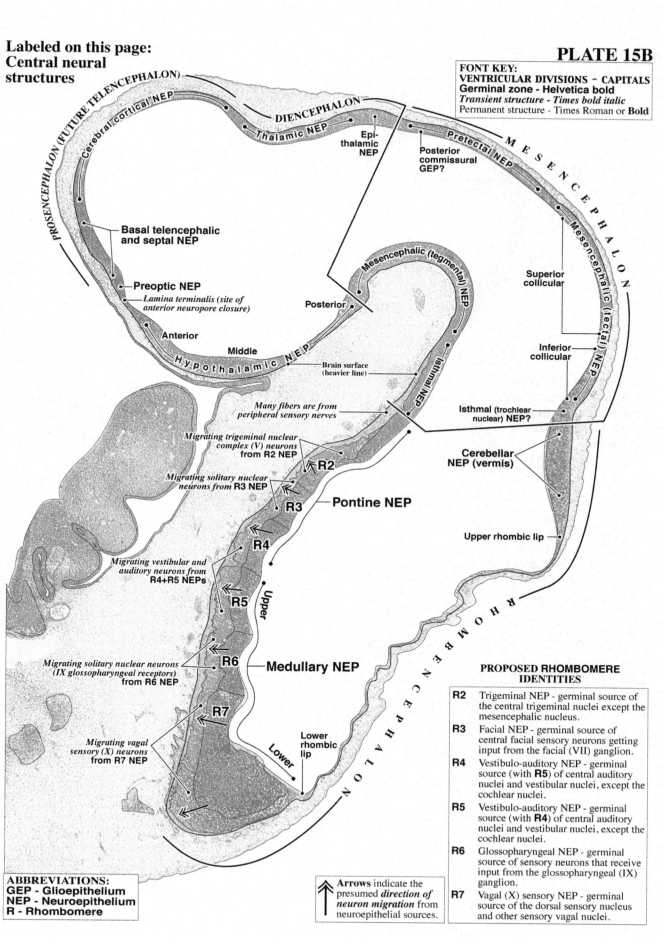

Labeled on this page:
Central neural
structures

PLATE 15B

FONT KEY:
VENTRICULAR DIVISIONS – CAPITALS
Germinal zone - Helvetica bold
Transient structure - Times bold italic
Permanent structure - Times Roman or **Bold**

PROSENCEPHALON (FUTURE TELENCEPHALON)

Cerebral cortical NEP

DIENCEPHALON

Thalamic NEP

Epi-
thalamic
NEP

Posterior
commissural
GEP?

Pretectal NEP

M E S E N C E P H A L O N

Mesencephalic (tectal) NEP

Basal telencephalic
and septal NEP

Preoptic NEP

*Lamina terminalis (site of
anterior neuropore closure)*

Anterior

Middle

H y p o t h a l a m i c NEP

Posterior

Mesencephalic (tegmental) NEP

Superior
collicular

Inferior
collicular

Brain surface
(heavier line)

Isthmal NEP

Isthmal (trochlear
nuclear) NEP?

*Many fibers are from
peripheral sensory nerves*

Cerebellar
NEP (vermis)

*Migrating trigeminal nuclear
complex (V) neurons
from R2 NEP*

R2

*Migrating solitary nuclear
neurons from R3 NEP*

R3

Pontine NEP

Upper rhombic lip

R4

*Migrating vestibular and
auditory neurons from
R4+R5 NEPs*

R5 Upper

R H O M B E N C E P H A L O N

*Migrating solitary nuclear neurons
(IX glossopharyngeal receptors)
from R6 NEP*

R6 Medullary NEP

R7

Lower
rhombic
lip

*Migrating vagal
sensory (X) neurons
from R7 NEP*

Lower

PROPOSED RHOMBOMERE
IDENTITIES

R2	Trigeminal NEP - germinal source of the central trigeminal nuclei except the mesencephalic nucleus.
R3	Facial NEP - germinal source of central facial sensory neurons getting input from the facial (VII) ganglion.
R4	Vestibulo-auditory NEP - germinal source (with **R5**) of central auditory nuclei and vestibular nuclei, except the cochlear nuclei.
R5	Vestibulo-auditory NEP - germinal source (with **R4**) of central auditory nuclei and vestibular nuclei, except the cochlear nuclei.
R6	Glossopharyngeal NEP - germinal source of sensory neurons that receive input from the glossopharyngeal (IX) ganglion.
R7	Vagal (X) sensory NEP - germinal source of the dorsal sensory nucleus and other sensory vagal nuclei.

ABBREVIATIONS:
GEP - Glioepithelium
NEP - Neuroepithelium
R - Rhombomere

Arrows indicate the
presumed *direction of
neuron migration* from
neuroepithelial sources.

PLATE 16A

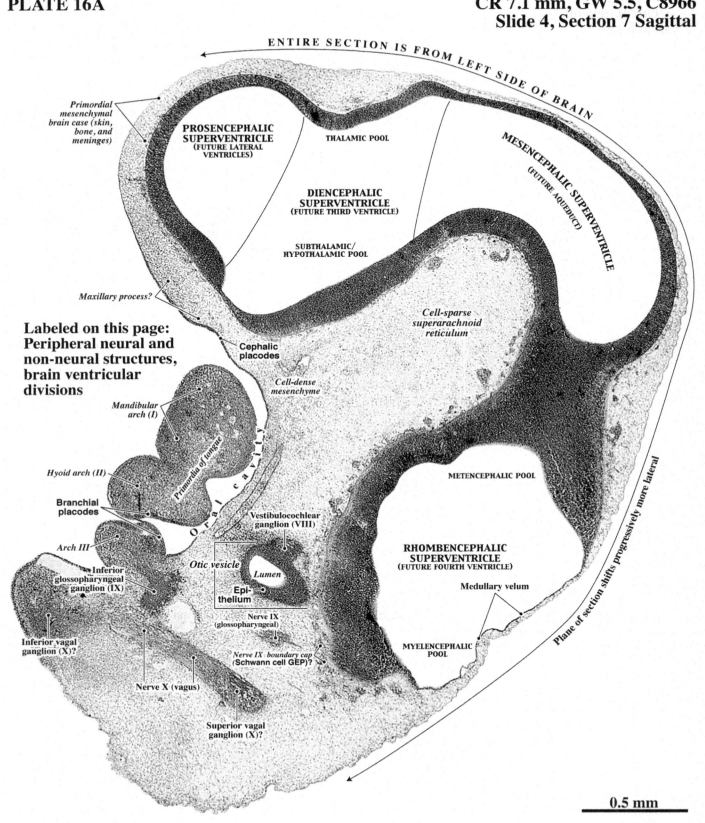

ENTIRE SECTION IS FROM LEFT SIDE OF BRAIN

Primordial mesenchymal brain case (skin, bone, and meninges)

PROSENCEPHALIC SUPERVENTRICLE
(FUTURE LATERAL VENTRICLES)

THALAMIC POOL

MESENCEPHALIC SUPERVENTRICLE
(FUTURE AQUEDUCT)

DIENCEPHALIC SUPERVENTRICLE
(FUTURE THIRD VENTRICLE)

SUBTHALAMIC/
HYPOTHALAMIC POOL

Cell-sparse superarachnoid reticulum

Maxillary process?

Labeled on this page:
Peripheral neural and non-neural structures, brain ventricular divisions

Cephalic placodes

Cell-dense mesenchyme

Mandibular arch (I)

Primordia of tongue

Oral cavity

METENCEPHALIC POOL

Hyoid arch (II)

Branchial placodes

Vestibulocochlear ganglion (VIII)

Arch III

Otic vesicle

Lumen

RHOMBENCEPHALIC SUPERVENTRICLE
(FUTURE FOURTH VENTRICLE)

Inferior glossopharyngeal ganglion (IX)

Epi-
thelium

Medullary velum

Nerve IX
(glossopharyngeal)

Inferior vagal ganglion (X)?

Nerve IX boundary cap (Schwann cell GEP)?

MYELENCEPHALIC POOL

Plane of section shifts progressively more lateral

Nerve X (vagus)

Superior vagal ganglion (X)?

0.5 mm

**Labeled on this page:
Central neural
structures**

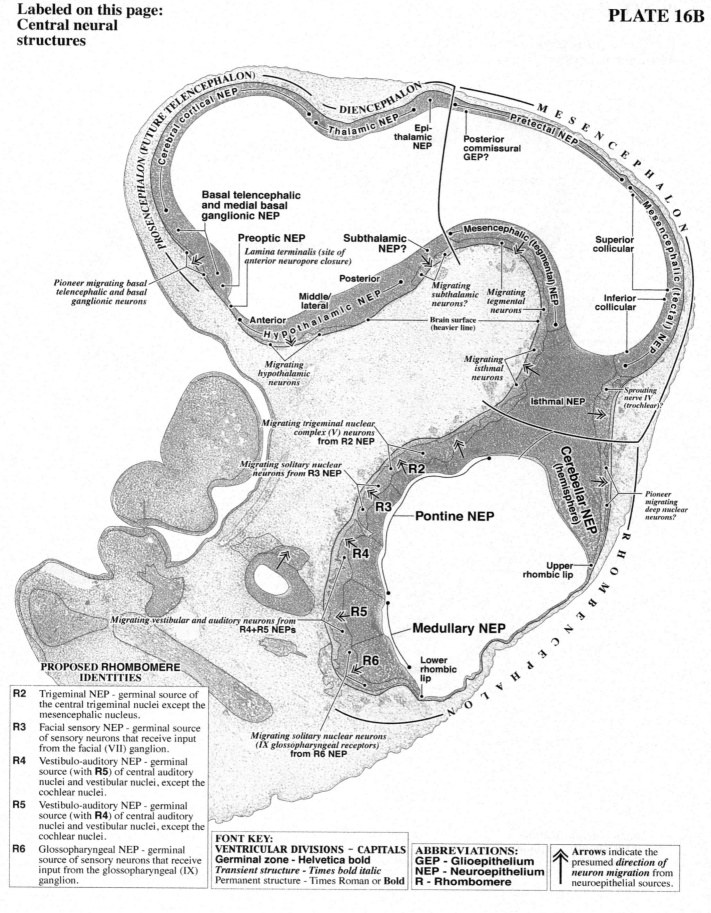

**PROPOSED RHOMBOMERE
IDENTITIES**

R2 Trigeminal NEP - germinal source of
the central trigeminal nuclei except the
mesencephalic nucleus.

R3 Facial sensory NEP - germinal source
of sensory neurons that receive input
from the facial (VII) ganglion.

R4 Vestibulo-auditory NEP - germinal
source (with **R5**) of central auditory
nuclei and vestibular nuclei, except the
cochlear nuclei.

R5 Vestibulo-auditory NEP - germinal
source (with **R4**) of central auditory
nuclei and vestibular nuclei, except the
cochlear nuclei.

R6 Glossopharyngeal NEP - germinal
source of sensory neurons that receive
input from the glossopharyngeal (IX)
ganglion.

**FONT KEY:
VENTRICULAR DIVISIONS – CAPITALS
Germinal zone - Helvetica bold
Transient structure - Times bold italic
Permanent structure - Times Roman or Bold**

**ABBREVIATIONS:
GEP - Glioepithelium
NEP - Neuroepithelium
R - Rhombomere**

Arrows indicate the
presumed *direction of
neuron migration* from
neuroepithelial sources.

Labeled on this page: Peripheral neural and
non-neural structures, brain ventricular divisions

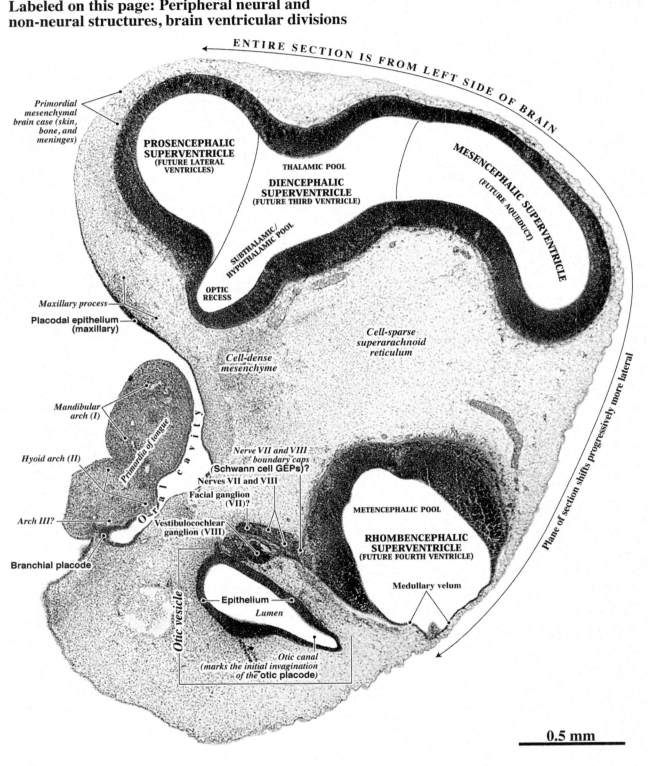

ENTIRE SECTION IS FROM LEFT SIDE OF BRAIN

*Primordial
mesenchymal
brain case (skin,
bone, and
meninges)*

**PROSENCEPHALIC
SUPERVENTRICLE
(FUTURE LATERAL
VENTRICLES)**

THALAMIC POOL

**DIENCEPHALIC
SUPERVENTRICLE
(FUTURE THIRD VENTRICLE)**

MESENCEPHALIC SUPERVENTRICLE
(FUTURE AQUEDUCT)

SUBTHALAMIC/
HYPOTHALAMIC POOL

OPTIC
RECESS

Maxillary process

**Placodal epithelium
(maxillary)**

*Cell-dense
mesenchyme*

*Cell-sparse
superarachnoid
reticulum*

*Mandibular
arch (I)*

Primordia of tongue

Hyoid arch (II)

*Nerve VII and VIII
boundary caps*
(Schwann cell GEPs)?

Nerves VII and VIII

**Facial ganglion
(VII)?**

Arch III?

Oral cavity

**Vestibulocochlear
ganglion (VIII)**

METENCEPHALIC POOL

**RHOMBENCEPHALIC
SUPERVENTRICLE
(FUTURE FOURTH VENTRICLE)**

Branchial placode

Otic vesicle

Epithelium

Lumen

Medullary velum

*Otic canal
(marks the initial invagination
of the otic placode)*

Plane of section shifts progressively more lateral

0.5 mm

**See a higher-magnification
view of the rhombencephalon
from this section in
Plates 22A and B.**

Labeled on this page:
Central neural
structures

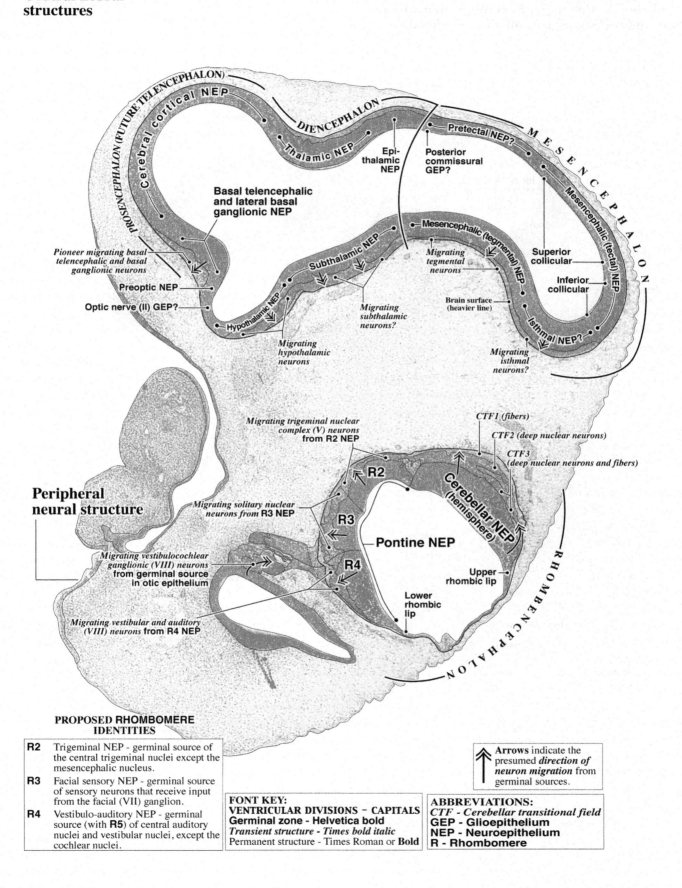

PROSENCEPHALON (FUTURE TELENCEPHALON)

Cerebral cortical NEP

DIENCEPHALON

Thalamic NEP

Epi-
thalamic
NEP

Posterior
commissural
GEP?

Pretectal NEP?

MESENCEPHALON

Mesencephalic (tectal) NEP

Basal telencephalic
and lateral basal
ganglionic NEP

Mesencephalic (tegmental) NEP

Pioneer migrating basal
telencephalic and basal
ganglionic neurons

Subthalamic NEP

Migrating
tegmental
neurons

Superior
collicular

Preoptic NEP

Inferior
collicular

Optic nerve (II) GEP?

Hypothalamic NEP

Migrating
subthalamic
neurons?

Brain surface
(heavier line)

Isthmal NEP?

Migrating
hypothalamic
neurons

Migrating
isthmal
neurons?

Migrating trigeminal nuclear
complex (V) neurons
from R2 NEP

CTF1 (fibers)

CTF2 (deep nuclear neurons)

CTF3
(deep nuclear neurons and fibers)

Peripheral
neural structure

R2

Cerebellar NEP
(hemisphere)

Migrating solitary nuclear
neurons from R3 NEP

R3

Pontine NEP

Migrating vestibulocochlear
ganglionic (VIII) neurons
from germinal source
in otic epithelium

R4

Upper
rhombic
lip

RHOMBENCEPHALON

Lower
rhombic
lip

Migrating vestibular and auditory
(VIII) neurons **from R4 NEP**

PROPOSED RHOMBOMERE
IDENTITIES

R2 Trigeminal NEP - germinal source of
the central trigeminal nuclei except the
mesencephalic nucleus.

R3 Facial sensory NEP - germinal source
of sensory neurons that receive input
from the facial (VII) ganglion.

R4 Vestibulo-auditory NEP - germinal
source (with **R5**) of central auditory
nuclei and vestibular nuclei, except the
cochlear nuclei.

Arrows indicate the
presumed *direction of*
neuron migration from
germinal sources.

FONT KEY:
VENTRICULAR DIVISIONS – CAPITALS
Germinal zone - Helvetica bold
Transient structure - Times bold italic
Permanent structure - Times Roman or **Bold**

ABBREVIATIONS:
CTF - Cerebellar transitional field
GEP - Glioepithelium
NEP - Neuroepithelium
R - Rhombomere

PLATE 18A

Labeled on this page: Peripheral neural and non-neural structures, brain ventricular divisions

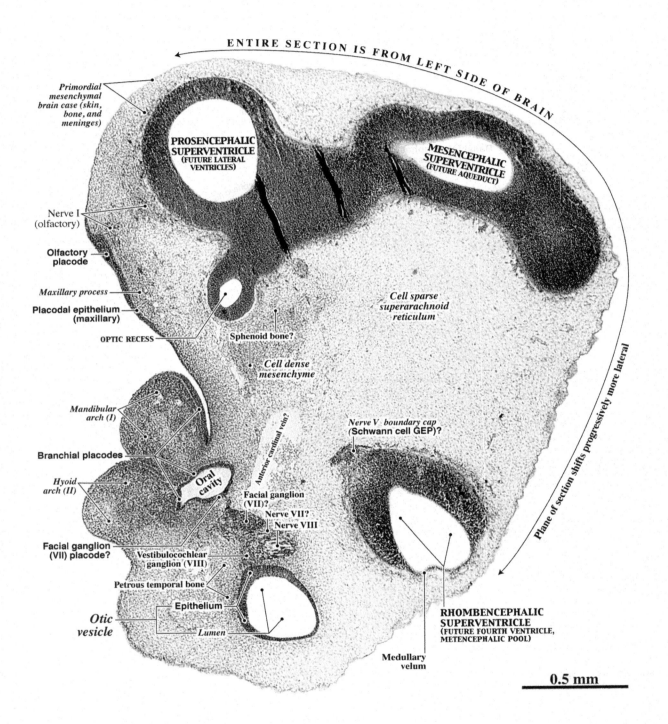

ENTIRE SECTION IS FROM LEFT SIDE OF BRAIN

Primordial mesenchymal brain case (skin, bone, and meninges)

PROSENCEPHALIC SUPERVENTRICLE (FUTURE LATERAL VENTRICLES)

MESENCEPHALIC SUPERVENTRICLE (FUTURE AQUEDUCT)

Nerve I (olfactory)

Olfactory placode

Maxillary process

Placodal epithelium (maxillary)

OPTIC RECESS

Sphenoid bone?

Cell sparse superarachnoid reticulum

Cell dense mesenchyme

Plane of section shifts progressively more lateral

Mandibular arch (I)

Anterior cardinal vein?

Nerve V boundary cap (Schwann cell GEP)?

Branchial placodes

Hyoid arch (II)

Oral cavity

Facial ganglion (VII)?

Nerve VII?
Nerve VIII

Facial ganglion (VII) placode?

Vestibulocochlear ganglion (VIII)

Petrous temporal bone

Epithelium

Otic vesicle

Lumen

RHOMBENCEPHALIC SUPERVENTRICLE (FUTURE FOURTH VENTRICLE, METENCEPHALIC POOL)

Medullary velum

0.5 mm

**Labeled on this page:
Central neural
structures**

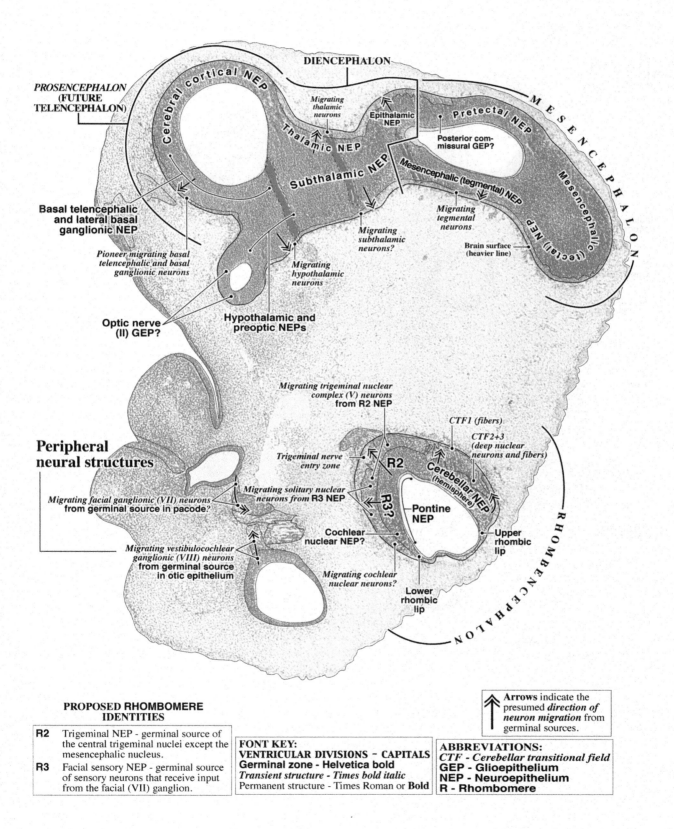

**Peripheral
neural structures**

**PROPOSED RHOMBOMERE
IDENTITIES**

R2 Trigeminal NEP - germinal source of
the central trigeminal nuclei except the
mesencephalic nucleus.

R3 Facial sensory NEP - germinal source
of sensory neurons that receive input
from the facial (VII) ganglion.

FONT KEY:
VENTRICULAR DIVISIONS – CAPITALS
Germinal zone - Helvetica bold
Transient structure - Times bold italic
Permanent structure - Times Roman or **Bold**

Arrows indicate the
presumed *direction of
neuron migration* from
germinal sources.

ABBREVIATIONS:
CTF - Cerebellar transitional field
GEP - Glioepithelium
NEP - Neuroepithelium
R - Rhombomere

PLATE 19A

Labeled on this page: Peripheral neural and non-neural structures, brain ventricular divisions

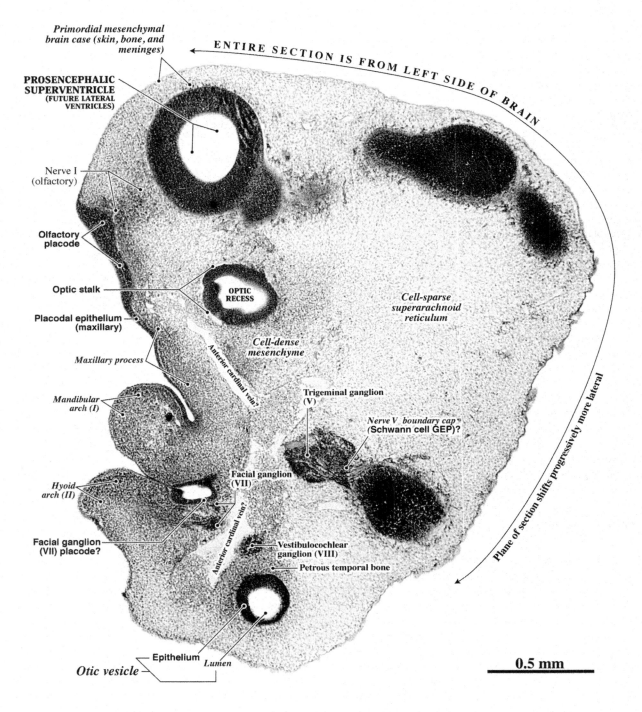

Primordial mesenchymal brain case (skin, bone, and meninges)

ENTIRE SECTION IS FROM LEFT SIDE OF BRAIN

PROSENCEPHALIC SUPERVENTRICLE (FUTURE LATERAL VENTRICLES)

Nerve I (olfactory)

Olfactory placode

Optic stalk

Placodal epithelium (maxillary)

Maxillary process

Mandibular arch (I)

Hyoid arch (II)

Facial ganglion (VII) placode?

Anterior cardinal vein?

OPTIC RECESS

Cell-dense mesenchyme

Cell-sparse superarachnoid reticulum

Trigeminal ganglion (V)

Nerve V boundary cap (Schwann cell GEP)?

Facial ganglion (VII)

Anterior cardinal vein?

Vestibulocochlear ganglion (VIII)

Petrous temporal bone

Plane of section shifts progressively more lateral

Epithelium *Lumen*

Otic vesicle

0.5 mm

Labeled on this page:
Central neural
structures

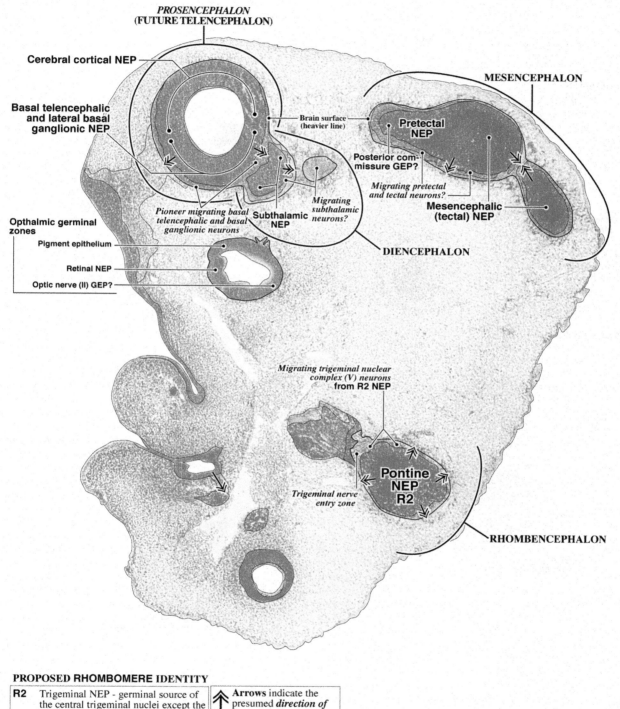

PROSENCEPHALON
(FUTURE TELENCEPHALON)

Cerebral cortical NEP

MESENCEPHALON

Basal telencephalic
and lateral basal
ganglionic NEP

Brain surface
(heavier line)

Pretectal
NEP

Posterior com-
missure GEP?

Migrating pretectal
and tectal neurons?

Migrating
subthalamic
neurons?

Mesencephalic
(tectal) NEP

Opthalmic germinal
zones

Pioneer migrating basal
telencephalic and basal
ganglionic neurons

Subthalamic
NEP

Pigment epithelium

DIENCEPHALON

Retinal NEP

Optic nerve (II) GEP?

Migrating trigeminal nuclear
complex (V) neurons
from R2 NEP

Pontine
NEP
R2

Trigeminal nerve
entry zone

RHOMBENCEPHALON

PROPOSED RHOMBOMERE IDENTITY

R2	Trigeminal NEP - germinal source of the central trigeminal nuclei except the mesencephalic nucleus.

Arrows indicate the presumed *direction of neuron migration* from germinal sources.

FONT KEY:
VENTRICULAR DIVISIONS – CAPITALS
Germinal zone - Helvetica bold
Transient structure - Times bold italic
Permanent structure - Times Roman or **Bold**

ABBREVIATIONS:
GEP - Glioepithelium
NEP - Neuroepithelium
R - Rhombomere

58

PLATE 20A

**Labeled on this page: Peripheral neural and
non-neural structures, brain ventricular divisions**

**CR 7.1 mm, GW 5.5, C8966
Slide 2, Section 22 Sagittal**

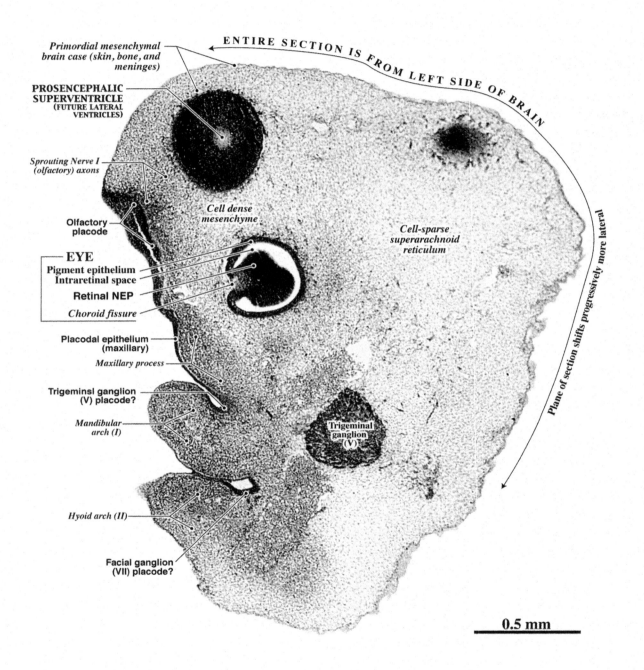

*Primordial mesenchymal
brain case (skin, bone, and
meninges)*

**PROSENCEPHALIC
SUPERVENTRICLE
(FUTURE LATERAL
VENTRICLES)**

*Sprouting Nerve I
(olfactory) axons*

**Olfactory
placode**

EYE
Pigment epithelium
Intraretinal space
Retinal NEP
Choroid fissure

**Placodal epithelium
(maxillary)**

Maxillary process

**Trigeminsl ganglion
(V) placode?**

*Mandibular
arch (I)*

Hyoid arch (II)

**Facial ganglion
(VII) placode?**

*Cell dense
mesenchyme*

*Cell-sparse
superarachnoid
reticulum*

ENTIRE SECTION IS FROM LEFT SIDE OF BRAIN

Plane of section shifts progressively more lateral

Trigeminal
ganglion
(V)

0.5 mm

**Labeled on this page:
Central neural
structures**

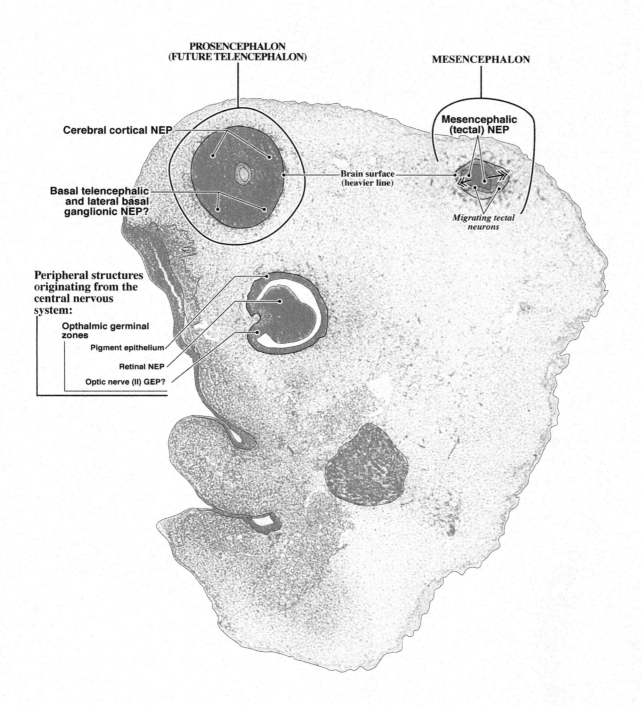

PROSENCEPHALON
(FUTURE TELENCEPHALON)

MESENCEPHALON

Cerebral cortical NEP

**Mesencephalic
(tectal) NEP**

Brain surface
(heavier line)

**Basal telencephalic
and lateral basal
ganglionic NEP?**

*Migrating tectal
neurons*

**Peripheral structures
originating from the
central nervous
system:**

**Opthalmic germinal
zones**

Pigment epithelium

Retinal NEP

Optic nerve (II) GEP?

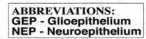

Arrows indicate the
presumed *direction of
neuron migration* from
germinal sources.

**ABBREVIATIONS:
GEP - Glioepithelium
NEP - Neuroepithelium**

FONT KEY:
**VENTRICULAR DIVISIONS – CAPITALS
Germinal zone - Helvetica bold
*Transient structure - Times bold italic***
Permanent structure - Times Roman or **Bold**

HYPOTHALAMUS, MESENCEPHALON, AND RHOMBENCEPHALON

PROPOSED RHOMBOMERE IDENTITIES

R2   Trigeminal NEP - germinal source of the central trigeminal nuclei except the mesencephalic nucleus.

R3   Facial sensory NEP - germinal source of sensory neurons that receive input from the facial (VII) ganglion.

R4   Vestibulo-auditory NEP - germinal source (with **R5**) of central auditory nuclei and vestibular nuclei, except the cochlear nuclei.

R5   Vestibulo-auditory NEP - germinal source (with **R4**) of central auditory nuclei and vestibular nuclei, except the cochlear nuclei.

R6   Glossopharyngeal NEP - germinal source of sensory neurons that receive input from the glossopharyngeal (IX) ganglion.

R7   Vagal (X) sensory NEP - germinal source of the dorsal sensory nucleus and other sensory vagal nuclei.

PLATE 21A
CR 7.1 mm, GW 5.5, C8966
Slide 5, Section 2 Sagittal

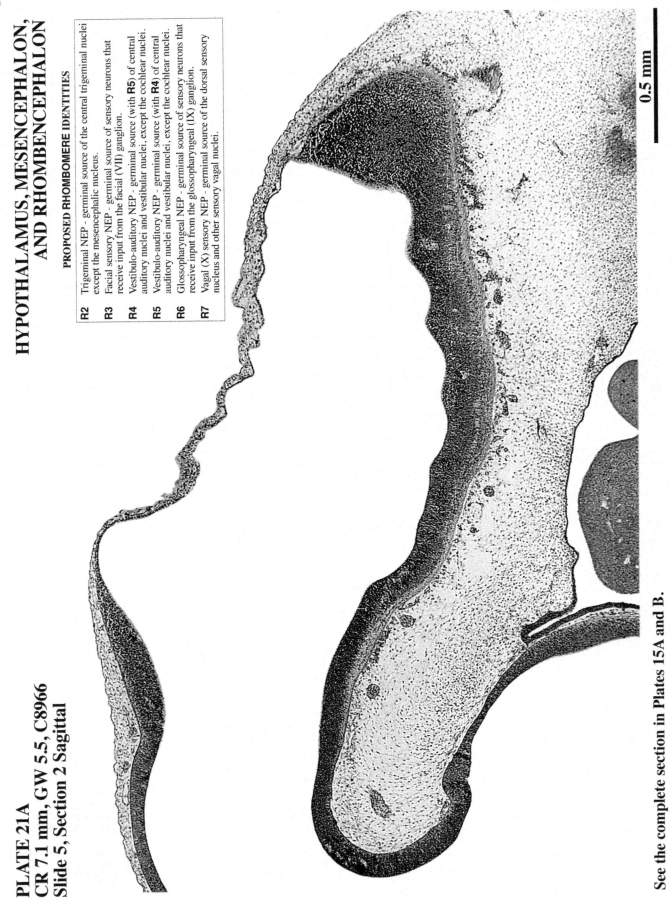

See the complete section in Plates 15A and B.

0.5 mm

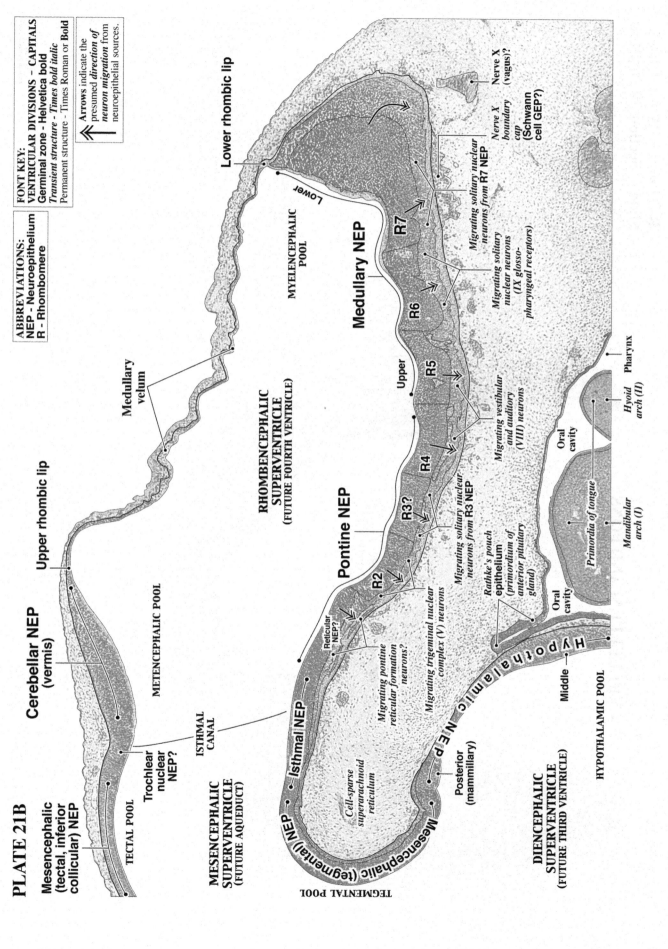

61

PLATE 21B

FONT KEY:
VENTRICULAR DIVISIONS – CAPITALS
Germinal zone - Helvetica bold
Transient structure - Times bold italic
Permanent structure - Times Roman or **Bold**

⟸ **Arrows** indicate the presumed *direction of neuron migration* from neuroepithelial sources.

ABBREVIATIONS:
NEP - Neuroepithelium
R - Rhombomere

Mesencephalic (tectal, inferior collicular) NEP

Cerebellar NEP (vermis)

Upper rhombic lip

TECTAL POOL

Medullary velum

Lower rhombic lip

Trochlear nuclear NEP?

METENCEPHALIC POOL

ISTHMAL CANAL

MYELENCEPHALIC POOL

Lower

Medullary NEP

RHOMBENCEPHALIC SUPERVENTRICLE (FUTURE FOURTH VENTRICLE)

MESENCEPHALIC SUPERVENTRICLE (FUTURE AQUEDUCT)

Isthmal NEP

Pontine NEP

R7

R6

Upper

R5

R4

R3?

R2

Reticular NEP?

Migrating pontine reticular formation neurons?

Migrating trigeminal nuclear complex (V) neurons?

Migrating solitary nuclear neurons from R3 NEP

Rathke's pouch epithelium (primordium of anterior pituitary gland)

Migrating solitary nuclear neurons from R7 NEP

Migrating solitary nuclear neurons (IX glosso-pharyngeal receptors)

Migrating vestibular and auditory (VIII) neurons

Nerve X (vagus)?

Nerve X boundary cap (Schwann cell GEP?)

Mesencephalic (tegmental) NEP

Cell-sparse superarachnoid reticulum

Posterior (mammillary)

H y p o t h a l a m i c   N E P

Middle

Oral cavity

Oral cavity

Rathke's pouch

Primordia of tongue

Mandibular arch (I)

Hyoid arch (II)

Pharynx

DIENCEPHALIC SUPERVENTRICLE (FUTURE THIRD VENTRICLE)

HYPOTHALAMIC POOL

TEGMENTAL POOL

62

PLATE 22A

CEREBELLUM AND PONS

CR 7.1 mm, GW 5.5, C8966
Slide 3, Section 24 Sagittal

0.1 mm

PROPOSED RHOMBOMERE
IDENTITIES

R2 Trigeminal NEP - germinal source of
the central trigeminal nuclei except the
mesencephalic nucleus.

R3 Facial sensory NEP - germinal source
of sensory neurons that receive input
from the facial (VII) ganglion.

R4 Vestibulo-auditory NEP - germinal
source (with R5) of central auditory
nuclei and vestibular nuclei, except the
cochlear nuclei.

See the entire section in Plates 17A and B.

63

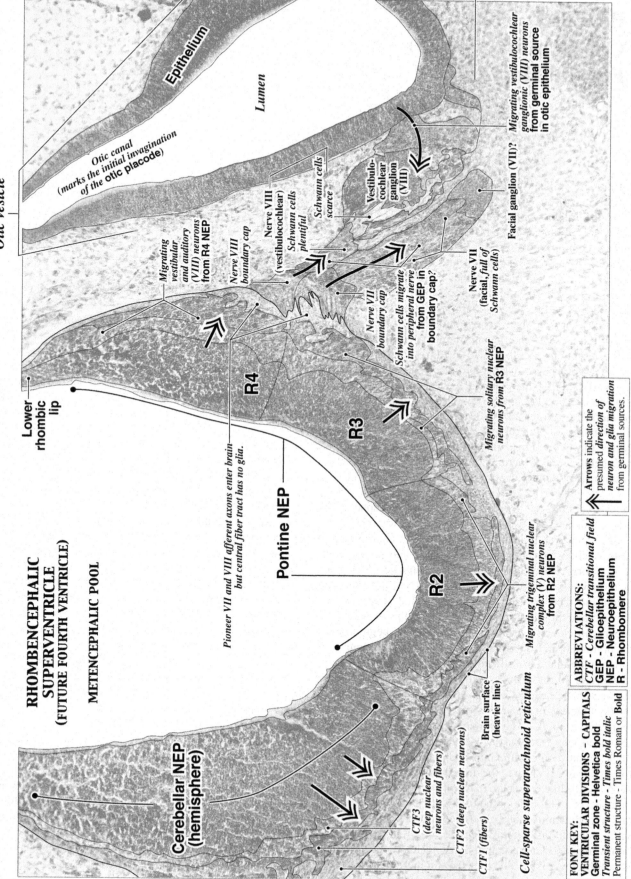

PLATE 22B

RHOMBENCEPHALIC SUPERVENTRICLE (FUTURE FOURTH VENTRICLE)

METENCEPHALIC POOL

Otic vesicle

Epithelium

Lumen

Otic canal (marks the initial invagination of the otic placode)

Lower rhombic lip

Migrating vestibulocochlear ganglionic (VIII) neurons from germinal source in otic epithelium

Cerebellar NEP (hemisphere)

Pioneer VII and VIII afferent axons enter brain but central fiber tract has no glia.

Migrating vestibular and auditory (VIII) neurons from R4 NEP

Nerve VIII boundary cap

Nerve VIII (vestibulocochlear)

Schwann cells plentiful

Schwann cells scarce

Vestibulo-cochlear ganglion (VIII)

Facial ganglion (VII)?

Nerve VII boundary cap

Nerve VII (facial, full of Schwann cells)

Schwann cells migrate into peripheral nerve from GEP in boundary cap?

R4

R3

R2

Pontine NEP

Migrating solitary nuclear neurons from R3 NEP

Migrating trigeminal nuclear complex (V) neurons from R2 NEP

Brain surface (heavier line)

Cell-sparse superarachnoid reticulum

CTF3 (deep nuclear neurons)

CTF2 (deep nuclear neurons and fibers)

CTF1 (fibers)

ABBREVIATIONS:
CTF - Cerebellar transitional field
GEP - Glioepithelium
NEP - Neuroepithelium
R - Rhombomere

Arrows indicate the presumed direction of neuron and glia migration from germinal sources.

FONT KEY:
VENTRICULAR DIVISIONS – CAPITALS
Germinal zone - Helvetica bold
Transient structure - Times bold italic
Permanent structure - Times Roman or Bold

PART IV: C8314
CR 8.0 mm (GW 6.0)
Frontal/Horizontal

Carnegie Collection specimen #8314 (designated here as C8314) with an 8.0-mm crown-rump length (CR) is estimated to be at gestational week (GW) 6.0. C8314 was fixed in formalin, embedded in a celloidin/paraffin mix, and was cut in 8-μm transverse sections that were stained with azan. Sections of the prosencephalon and anterior mesencephalon are cut in the coronal plane, but the plane shifts to predominantly horizontal in the posterior mesencephalon, pons and medulla. We photographed 39 sections at low magnification from the frontal prominence to the posterior tips of the mesencephalon and rhombencephalon. Eleven of these sections are illustrated in **Plates 23AB to 32AB**. All photographs were used to produce computer-aided 3-D reconstructions of the external features of C8314's brain and eye (**Figure 12**), and to show each illustrated section *in situ* (insets, **Plates 23A to 32A**). Each illustrated section shows the brain with all surrounding tissues. Labels in **A Plates** (normal-contrast images) identify non-neural and peripheral neural structures; labels in **B Plates** (low-contrast images) identify central neural structures.

At this stage of development, the prosencephalon is still a unicameral vesicle. The most anterior brain sections are tentatively identified as the future telencephalon, while sections through the optic vesicle and posterior to it are more clearly identified as diencephalic. All parts of the prosencephalic neuroepithelium (NEP) are rapidly increasing their pool of neuronal and glial stem cells as they expand the shorelines of the enlarging prosencephalic superventricle. But neurogenesis is already beginning in the brain floor in front of the optic vesicle; cell migration is virtually absent. The olfactory placode has not yet invaginated and forms a thick epithelium in the anterolateral surface of the head. Cell densities deep to the trajectory of the olfactory nerve are visible, but it is too early to call them nerve fibers. The evaginated optic vesicle forms a C-shaped curve around the developing lens, defining an thick retinal neuroepithelium and a thin pigment epithelium. The eye is close to the diencephalon and does not yet form a stalk-like extension that will eventually become the optic nerve. Throughout the future epithalamus, thalamus, and hypothalamus, neurogenesis is just beginning in only three populations: lateral preoptic area, lateral hypothalamic area, and lateral mammillary nucleus. A primordial plexiform layer is outside the NEPs in these brain areas.

The mesencephalon contains a stockbuilding NEP in the pretectum and tectum. The tegmental and isthmal NEPs are also stockbuilding but some parts are entering the neurogenetic phase. Some pioneer neurons destined to settle in the oculomotor complex are sequestered in the NEP. The subpial fiber band is very thin in the tegmentum, but thickens slightly in the isthmus as more fibers move in via sensory nerves and perhaps a few from the spinal cord.

Almost all the rhombomere NEPs are in the neurogenetic phase. Blood islands and sprouting pioneer axons form clefts in between them. A definite feathered basal layer lines the border of each rhombomere NEP as postmitotic neurons accumulate (sequester) there before migration. Trigeminal sensory neurons will start generation in the next phase, but the trigeminal motor nucleus is having its peak day—possibly some of the motor neurons are accumulating in the region were trigeminal sensory axons will enter. Much neurogenesis is taking place in the auditory and vestibular nuclei that will settle in the pons and medulla; these neurons may be sequestered in basal parts of the R4 and R5 NEPs. Neurogenesis of the solitary nucleus (taste processing) is beginning in earnest, and anterior neurons may be sequestering in the R3 NEP (for inputs from the facial nerve), posterior neurons in the NEPs of R4 and R5 (for inputs from the glossopharyngeal and vagal nerves). The subpial fiber band is thicker where sensory afferents enter the brain and the central sensory neurons that will interact with peripheral axons may be waiting for the right axons to come in before they migrate. Medial pontine and medullary NEPs are thinning because much neurogenesis has finished in the motor nuclei of the medulla and pons, the reticular formation, and some lateral raphe nuclei; young neurons have moved out leaving behind a NEP that has a depleted population of neural stem cells.

The primordial cerebellar NEP is identifiable in the most posterior sections of the rhombencephalon, behind the posterior tectal NEP. The NEP is in a robust stockbuilding stage because neurogenesis has not yet begun. The precerebellar NEP in the dorsal medulla is just entering the neurogenetic phase with the generation of a few external cuneate nuclear neurons. It is still too early to see any sequestration, let alone migration from the precerebellar nuclear NEP.

C8314 Computer-aided 3-D Brain Reconstructions

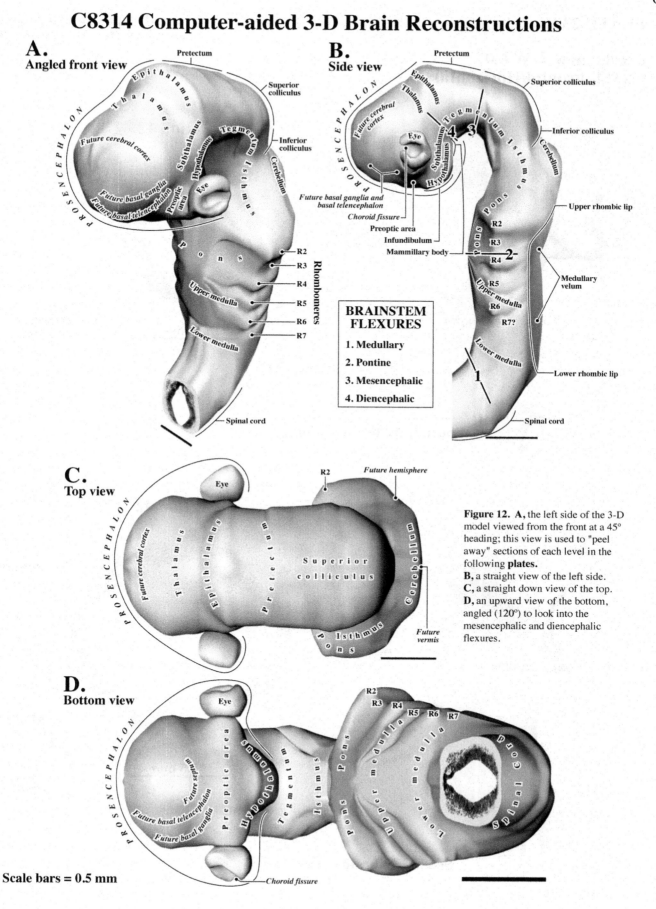

A. Angled front view

Pretectum
Superior colliculus
Inferior colliculus
Epithalamus
Thalamus
Future cerebral cortex
Subthalamus
Hypothalamus
Tegmentum
Isthmus
Cerebellum
Future basal ganglia
Future basal telencephalon
Preoptic area
Eye
PROSENCEPHALON
Pons
R2
R3
R4
R5
R6
R7
Rhombomeres
Upper medulla
Lower medulla
Spinal cord

B. Side view

Pretectum
Superior colliculus
Epithalamus
Thalamus
Future cerebral cortex
Eye
Tegmentum
Isthmus
Inferior colliculus
Cerebellum
Subthalamus
Hypothalamus
Future basal ganglia and basal telencephalon
Choroid fissure
Preoptic area
Infundibulum
Mammillary body
PROSENCEPHALON
Pons
R2
R3
R4
R5
R6
R7?
Upper medulla
Upper rhombic lip
Medullary velum
Lower medulla
Lower rhombic lip
Spinal cord

BRAINSTEM FLEXURES
1. Medullary
2. Pontine
3. Mesencephalic
4. Diencephalic

C. Top view

Eye
R2
Future hemisphere
PROSENCEPHALON
Future cerebral cortex
Thalamus
Epithalamus
Pretectum
Superior colliculus
Cerebellum
Isthmus
Pons
Future vermis

Figure 12. A, the left side of the 3-D model viewed from the front at a 45° heading; this view is used to "peel away" sections of each level in the following **plates.**
B, a straight view of the left side.
C, a straight down view of the top.
D, an upward view of the bottom, angled (120°) to look into the mesencephalic and diencephalic flexures.

D. Bottom view

Eye
R2
R3
R4
R5
R6
R7
PROSENCEPHALON
Future septum
Future basal telencephalon
Future basal ganglia
Preoptic area
Hypothalamus
Tegmentum
Isthmus
Pons
Upper medulla
Lower medulla
Spinal cord
Choroid fissure

Scale bars = 0.5 mm

PLATE 23A

CR 8.0 mm, GW 6.0
C8314, Frontal/Horizontal

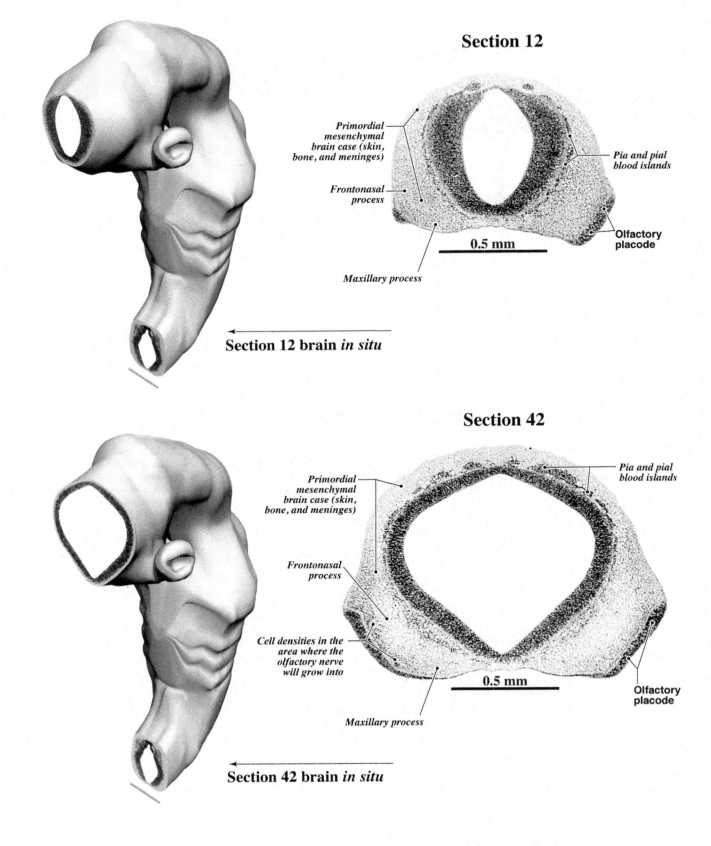

Section 12

*Primordial
mesenchymal
brain case (skin,
bone, and meninges)*

*Frontonasal
process*

*Pia and pial
blood islands*

Olfactory
placode

0.5 mm

Maxillary process

Section 12 brain *in situ*

Section 42

*Primordial
mesenchymal
brain case (skin,
bone, and meninges)*

*Frontonasal
process*

*Cell densities in the
area where the
olfactory nerve
will grow into*

*Pia and pial
blood islands*

0.5 mm

Olfactory
placode

Maxillary process

Section 42 brain *in situ*

Section 12

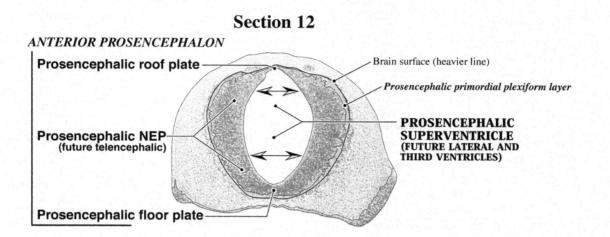

ANTERIOR PROSENCEPHALON

Prosencephalic roof plate

Brain surface (heavier line)

Prosencephalic primordial plexiform layer

Prosencephalic NEP
(future telencephalic)

PROSENCEPHALIC
SUPERVENTRICLE
(FUTURE LATERAL AND
THIRD VENTRICLES)

Prosencephalic floor plate

Section 42

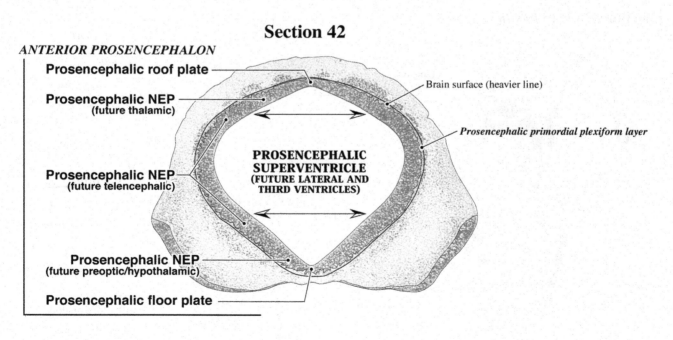

ANTERIOR PROSENCEPHALON

Prosencephalic roof plate

Brain surface (heavier line)

Prosencephalic NEP
(future thalamic)

Prosencephalic primordial plexiform layer

Prosencephalic NEP
(future telencephalic)

PROSENCEPHALIC
SUPERVENTRICLE
(FUTURE LATERAL AND
THIRD VENTRICLES)

Prosencephalic NEP
(future preoptic/hypothalamic)

Prosencephalic floor plate

NEP - Neuroepithelium

FONT KEY:
VENTRICULAR DIVISIONS - CAPITALS
Germinal zone - Helvetica bold
Transient structure - Times bold italic
Permanent structure - Times Roman or **Bold**

Arrows indicate the regionally
expanding shoreline of the
superventricle with increase in
stockbuilding NEP cells.

PLATE 24A

CR 8.0 mm, GW 6.0
C8314, Frontal/Horizontal
Section 82

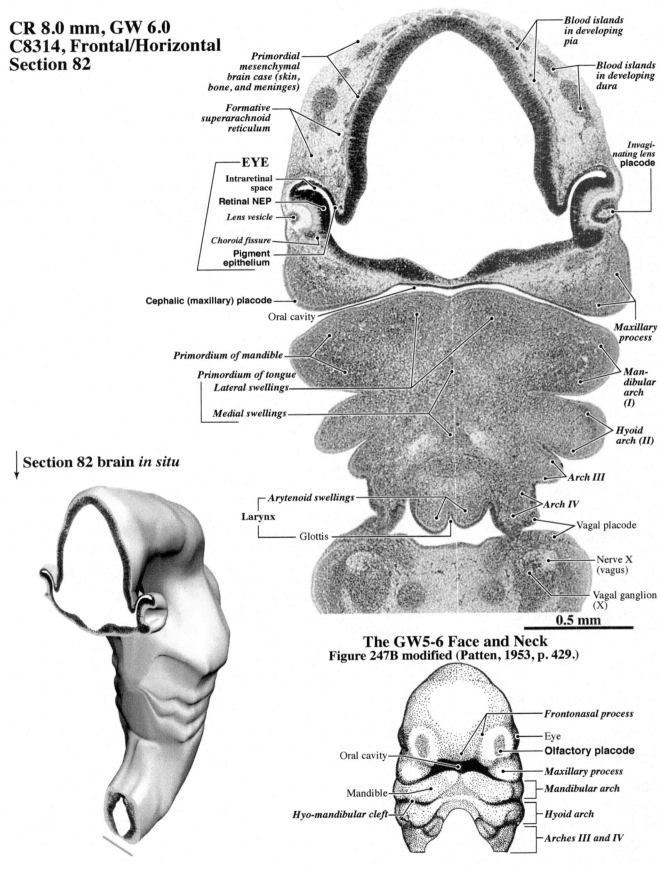

Blood islands in developing pia

Blood islands in developing dura

Primordial mesenchymal brain case (skin, bone, and meninges)

Formative superarachnoid reticulum

Invaginating lens placode

EYE

Intraretinal space

Retinal NEP

Lens vesicle

Choroid fissure

Pigment epithelium

Cephalic (maxillary) placode

Oral cavity

Maxillary process

Primordium of mandible

Primordium of tongue

Lateral swellings

Medial swellings

Mandibular arch (I)

Hyoid arch (II)

Arch III

Arch IV

Vagal placode

Section 82 brain *in situ*

Arytenoid swellings

Larynx

Glottis

Nerve X (vagus)

Vagal ganglion (X)

0.5 mm

The GW5-6 Face and Neck
Figure 247B modified (Patten, 1953, p. 429.)

Frontonasal process

Eye

Olfactory placode

Oral cavity

Maxillary process

Mandible

Mandibular arch

Hyo-mandibular cleft

Hyoid arch

Arches III and IV

Central neural structures labeled

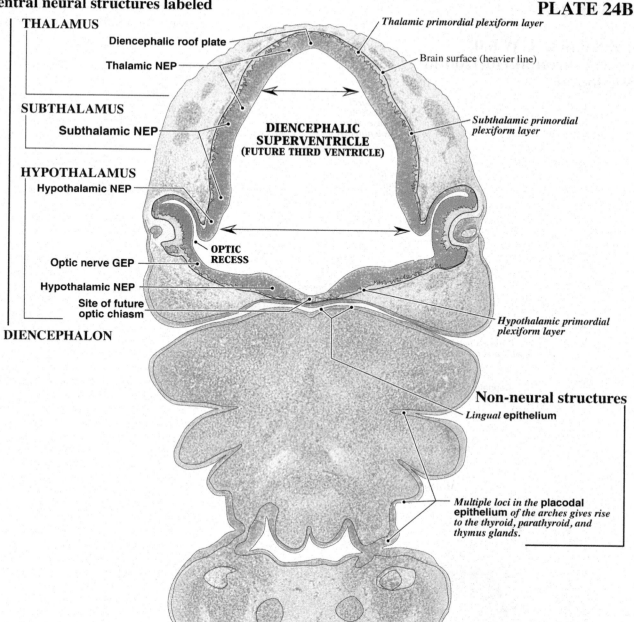

THALAMUS

Diencephalic roof plate

Thalamic NEP

Thalamic primordial plexiform layer

Brain surface (heavier line)

SUBTHALAMUS

Subthalamic NEP

Subthalamic primordial plexiform layer

DIENCEPHALIC SUPERVENTRICLE (FUTURE THIRD VENTRICLE)

HYPOTHALAMUS

Hypothalamic NEP

Optic nerve GEP

OPTIC RECESS

Hypothalamic NEP

Site of future optic chiasm

Hypothalamic primordial plexiform layer

DIENCEPHALON

Non-neural structures

Lingual epithelium

Multiple loci in the placodal epithelium *of the arches gives rise to the thyroid, parathyroid, and thymus glands.*

FONT KEY:
VENTRICULAR DIVISIONS - CAPITALS
Germinal zone - Helvetica bold
Transient structure - Times bold italic
Permanent structure - Times Roman or **Bold**

ABBREVIATIONS:
GEP - Glioepithelium
NEP - Neuroepithelium

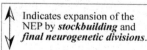

Indicates expansion of the NEP by *stockbuilding* and *final neurogenetic divisions*.

70

PLATE 25A

CR 8.0 mm, GW 6.0
C8314, Frontal/Horizontal
Section 97

Peripheral neural and non-neural structures labeled

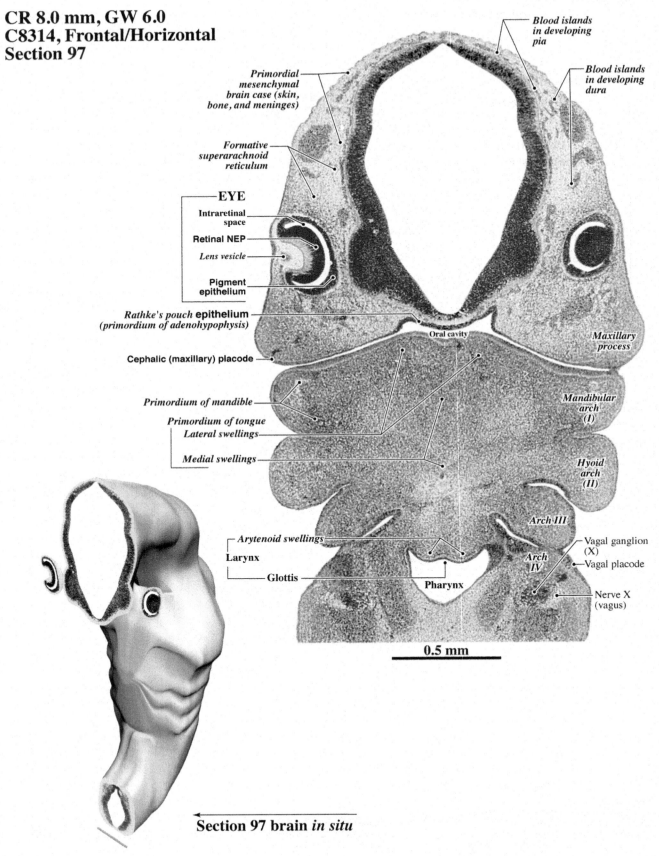

Blood islands in developing pia

Blood islands in developing dura

Primordial mesenchymal brain case (skin, bone, and meninges)

Formative superarachnoid reticulum

EYE

Intraretinal space

Retinal NEP

Lens vesicle

Pigment epithelium

Rathke's pouch epithelium (primordium of adenohypophysis)

Oral cavity

Maxillary process

Cephalic (maxillary) placode

Primordium of mandible

Primordium of tongue

Lateral swellings

Medial swellings

Mandibular arch (I)

Hyoid arch (II)

Arch III

Arytenoid swellings

Larynx

Glottis

Arch IV

Pharynx

Vagal ganglion (X)

Vagal placode

Nerve X (vagus)

0.5 mm

Section 97 brain in situ

Central neural structures labeled

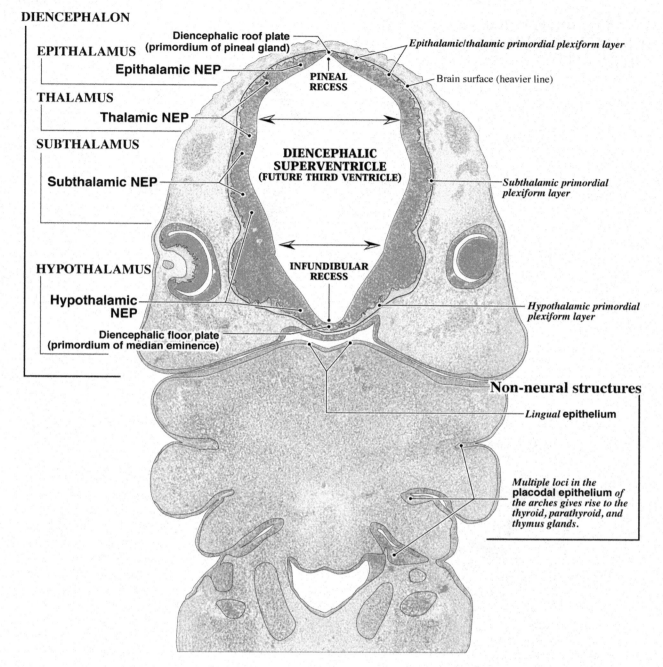

DIENCEPHALON

EPITHALAMUS

Diencephalic roof plate
(primordium of pineal gland)

Epithalamic/thalamic primordial plexiform layer

Epithalamic NEP

PINEAL
RECESS

Brain surface (heavier line)

THALAMUS

Thalamic NEP

SUBTHALAMUS

**DIENCEPHALIC
SUPERVENTRICLE
(FUTURE THIRD VENTRICLE)**

Subthalamic NEP

*Subthalamic primordial
plexiform layer*

INFUNDIBULAR
RECESS

HYPOTHALAMUS

**Hypothalamic
NEP**

*Hypothalamic primordial
plexiform layer*

Diencephalic floor plate
(primordium of median eminence)

Non-neural structures

Lingual epithelium

Multiple loci in the
placodal epithelium *of
the arches gives rise to the
thyroid, parathyroid, and
thymus glands.*

FONT KEY:
VENTRICULAR DIVISIONS - CAPITALS
Germinal zone - Helvetica bold
Transient structure - Times bold italic
Permanent structure - Times Roman or **Bold**

NEP - Neuroepithelium

↕ Indicates expansion of the
NEP by *stockbuilding* and
final neurogenetic divisions.

PLATE 26A

**CR 8.0 mm, GW 6.0
C8314, Frontal/Horizontal
Section 117**

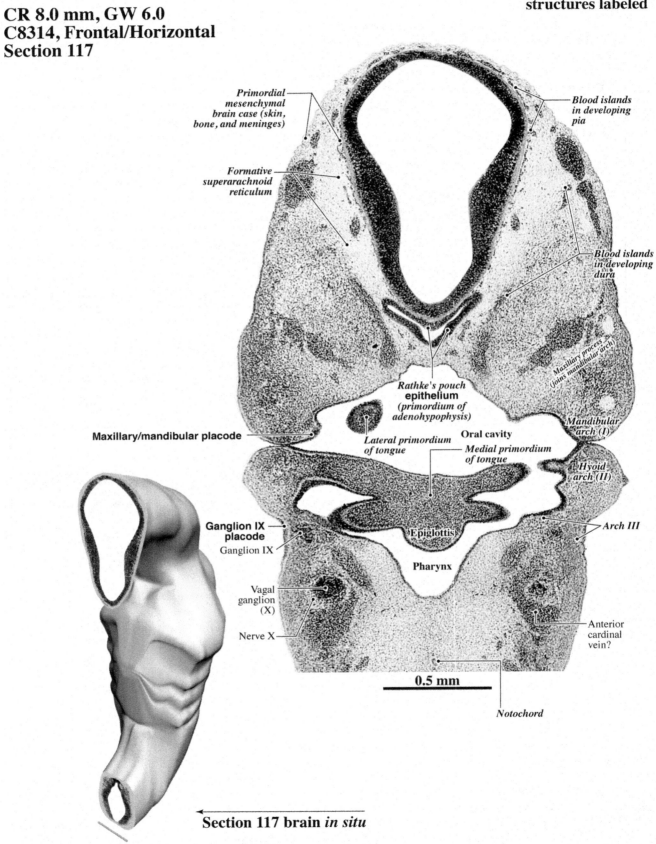

*Primordial
mesenchymal
brain case (skin,
bone, and meninges)*

*Blood islands
in developing
pia*

*Formative
superarachnoid
reticulum*

*Blood islands
in developing
dura*

*Maxillary process
(joins mandibular arch)*

Rathke's pouch
epithelium
*(primordium of
adenohypophysis)*

Oral cavity

*Mandibular
arch (I)*

Maxillary/mandibular placode

*Lateral primordium
of tongue*

*Medial primordium
of tongue*

*Hyoid
arch (II)*

**Ganglion IX
placode**

Ganglion IX

Epiglottis

Arch III

Pharynx

Vagal
ganglion
(X)

Nerve X

Anterior
cardinal
vein?

0.5 mm

Notochord

Section 117 brain *in situ*

MESENCEPHALON

Mesencephalic roof plate
(posterior commissural GEP)

MESENCEPHALIC SUPERVENTRICLE
(FUTURE AQUEDUCT)

PRETECTUM

Pretectal NEP

Pretectal/thalamic primordial
plexiform layer

Brain surface (heavier line)

THALAMUS

Thalamic NEP

DIENCEPHALIC
SUPERVENTRICLE
(FUTURE THIRD
VENTRICLE)

SUBTHALAMUS

Subthalamic NEP

Subthalamic primordial
plexiform layer

HYPOTHALAMUS

Hypothalamic
NEP

INFUNDI-
BULAR
RECESS

Hypothalamic primordial
plexiform layer

Diencephalic floor plate
(primordium of median eminence
and neurohypophysis)

DIENCEPHALON

Non-neural
structure

Multiple loci in the **placodal**
epithelium *of the arches*
gives rise to the thyroid,
parathyroid, and thymus
glands.

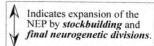

PLATE 27A

CR 8.0 mm, GW 6.0
C8314, Frontal/Horizontal
Section 127

Peripheral neural
and non-neural
structures labeled

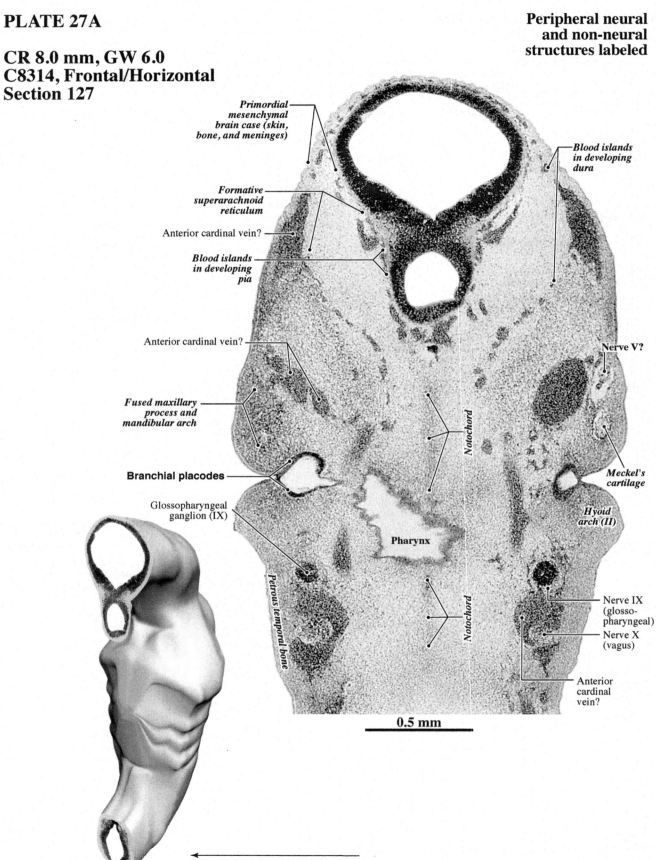

Primordial mesenchymal brain case (skin, bone, and meninges)

Formative superarachnoid reticulum

Anterior cardinal vein?

Blood islands in developing pia

Anterior cardinal vein?

Fused maxillary process and mandibular arch

Branchial placodes

Glossopharyngeal ganglion (IX)

Petrous temporal bone

Blood islands in developing dura

Nerve V?

Meckel's cartilage

Hyoid arch (II)

Notochord

Pharynx

Notochord

Nerve IX (glosso-pharyngeal)

Nerve X (vagus)

Anterior cardinal vein?

0.5 mm

Section 127 brain *in situ*

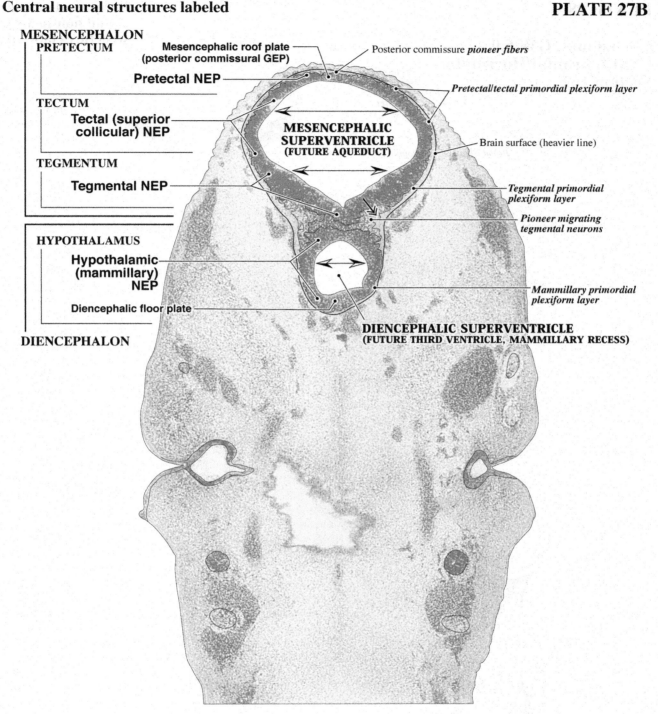

MESENCEPHALON
  PRETECTUM

  Mesencephalic roof plate — (posterior commissural GEP)

  Posterior commissure *pioneer fibers*

  Pretectal NEP

  Pretectal/tectal primordial plexiform layer

  TECTUM

  Tectal (superior collicular) NEP

  MESENCEPHALIC SUPERVENTRICLE (FUTURE AQUEDUCT)

  Brain surface (heavier line)

  TEGMENTUM

  Tegmental NEP

  Tegmental primordial plexiform layer

  Pioneer migrating tegmental neurons

  HYPOTHALAMUS

  Hypothalamic (mammillary) NEP

  Mammillary primordial plexiform layer

  Diencephalic floor plate

  DIENCEPHALIC SUPERVENTRICLE (FUTURE THIRD VENTRICLE, MAMMILLARY RECESS)

DIENCEPHALON

ABBREVIATIONS:
GEP - Glioepithelium
NEP - Neuroepithelium

FONT KEY:
VENTRICULAR DIVISIONS - CAPITALS
Germinal zone - Helvetica bold
Transient structure - Times bold italic
Permanent structure - Times Roman or **Bold**

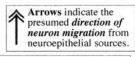

Arrows indicate the presumed *direction of neuron migration* from neuroepithelial sources.

Indicates expansion of the NEP by *stockbuilding* and *final neurogenetic divisions*.

76

PLATE 28A

CR 8.0 mm, GW 6.0
C8314, Frontal/Horizontal
Section 162

**Peripheral neural
and non-neural
structures labeled**

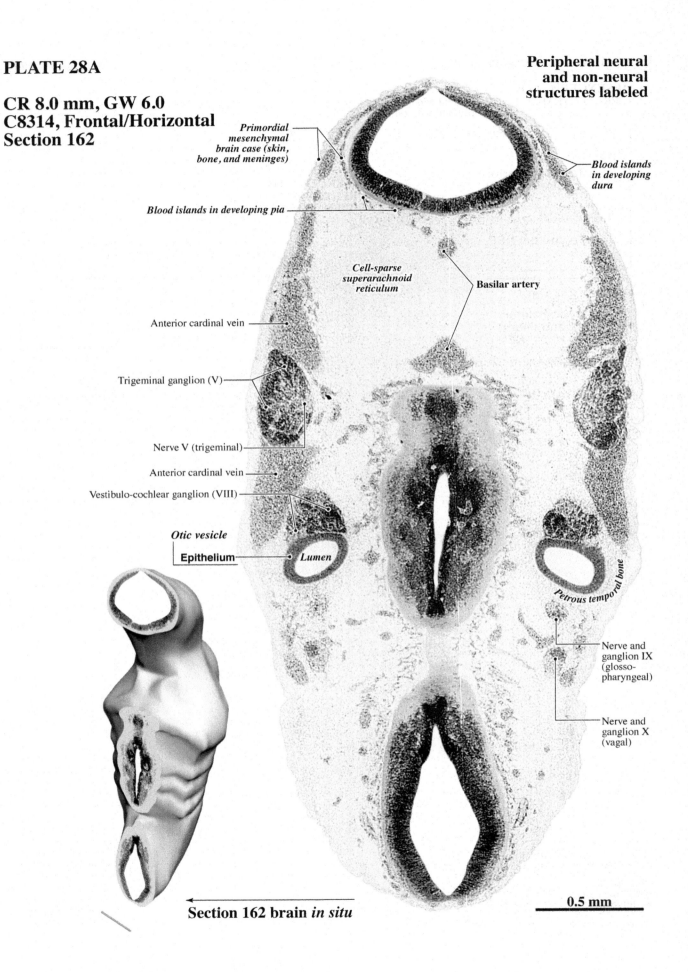

*Primordial
mesenchymal
brain case (skin,
bone, and meninges)*

*Blood islands
in developing
dura*

Blood islands in developing pia

*Cell-sparse
superarachnoid
reticulum*

Basilar artery

Anterior cardinal vein

Trigeminal ganglion (V)

Nerve V (trigeminal)

Anterior cardinal vein

Vestibulo-cochlear ganglion (VIII)

Otic vesicle
Epithelium
Lumen

Petrous temporal bone

Nerve and
ganglion IX
(glosso-
pharyngeal)

Nerve and
ganglion X
(vagal)

Section 162 brain *in situ*

0.5 mm

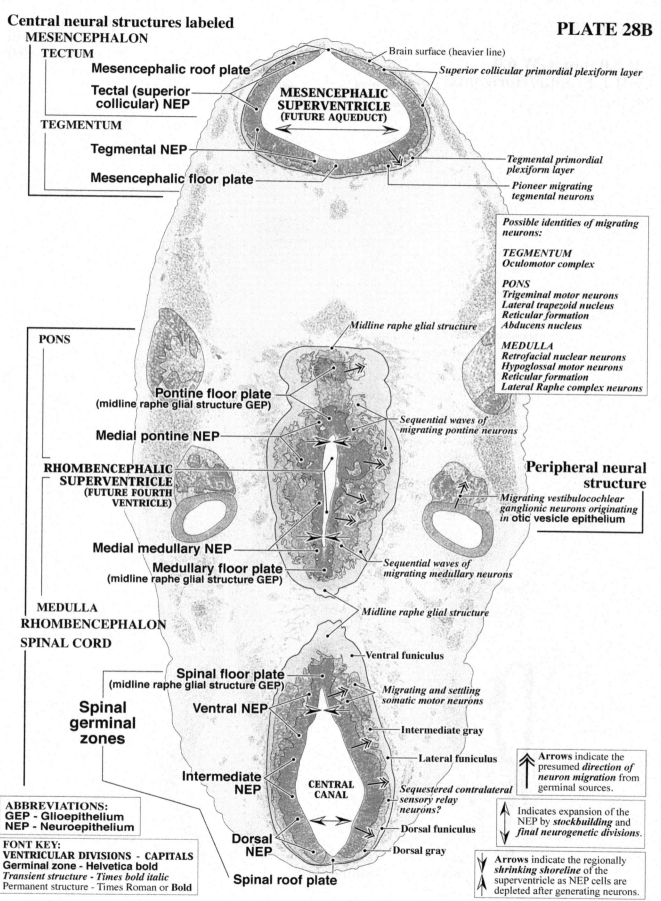

Central neural structures labeled

PLATE 28B

MESENCEPHALON
TECTUM
Mesencephalic roof plate
Tectal (superior collicular) NEP
TEGMENTUM
Tegmental NEP
Mesencephalic floor plate

Brain surface (heavier line)
Superior collicular primordial plexiform layer

MESENCEPHALIC SUPERVENTRICLE (FUTURE AQUEDUCT)

Tegmental primordial plexiform layer
Pioneer migrating tegmental neurons

Possible identities of migrating neurons:

TEGMENTUM
Oculomotor complex

PONS
Trigeminal motor neurons
Lateral trapezoid nucleus
Reticular formation
Abducens nucleus

MEDULLA
Retrofacial nuclear neurons
Hypoglossal motor neurons
Reticular formation
Lateral Raphe complex neurons

PONS

Midline raphe glial structure

Pontine floor plate
(midline raphe glial structure GEP)
Medial pontine NEP

RHOMBENCEPHALIC SUPERVENTRICLE (FUTURE FOURTH VENTRICLE)

Sequential waves of migrating pontine neurons

Peripheral neural structure

Migrating vestibulocochlear ganglionic neurons originating in otic vesicle epithelium

Medial medullary NEP
Medullary floor plate
(midline raphe glial structure GEP)

Sequential waves of migrating medullary neurons

MEDULLA
RHOMBENCEPHALON
SPINAL CORD

Midline raphe glial structure

Ventral funiculus

Spinal germinal zones

Spinal floor plate
(midline raphe glial structure GEP)
Ventral NEP

Migrating and settling somatic motor neurons

Intermediate gray

Lateral funiculus

Intermediate NEP

CENTRAL CANAL

Sequestered contralateral sensory relay neurons?

Dorsal funiculus

Dorsal NEP

Dorsal gray

Spinal roof plate

ABBREVIATIONS:
GEP - Glioepithelium
NEP - Neuroepithelium

FONT KEY:
VENTRICULAR DIVISIONS - CAPITALS
Germinal zone - Helvetica bold
Transient structure - Times bold italic
Permanent structure - Times Roman or **Bold**

Arrows indicate the presumed *direction of neuron migration* from germinal sources.

Indicates expansion of the NEP by *stockbuilding* and *final neurogenetic divisions*.

Arrows indicate the regionally *shrinking shoreline* of the superventricle as NEP cells are depleted after generating neurons.

PLATE 29A

CR 8.0 mm, GW 6.0
C8314, Frontal/Horizontal
Section 172

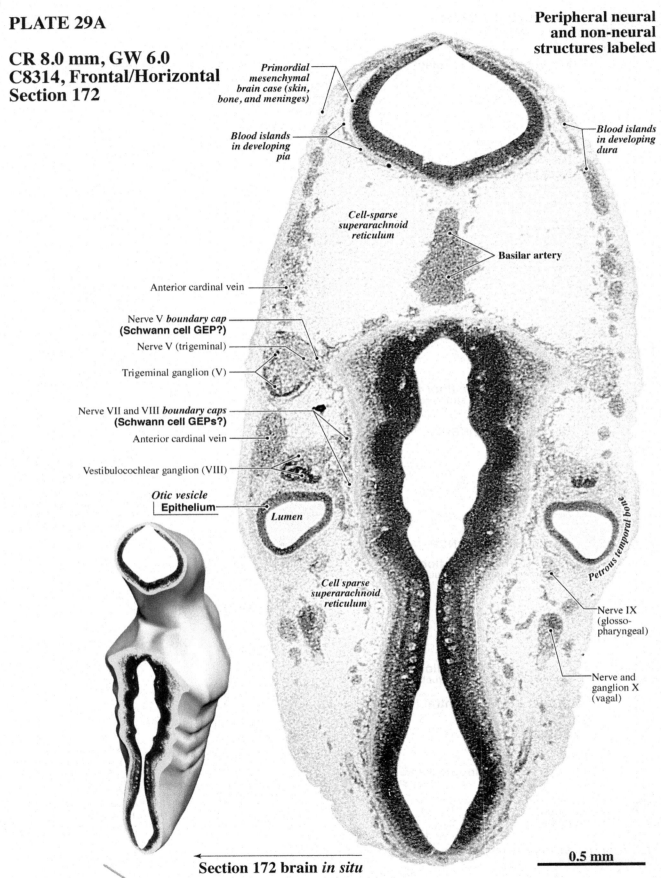

*Primordial
mesenchymal
brain case (skin,
bone, and meninges)*

*Blood islands
in developing
pia*

*Blood islands
in developing
dura*

*Cell-sparse
superarachnoid
reticulum*

Basilar artery

Anterior cardinal vein

Nerve V *boundary cap*
(Schwann cell GEP?)

Nerve V (trigeminal)

Trigeminal ganglion (V)

Nerve VII and VIII *boundary caps*
(Schwann cell GEPs?)

Anterior cardinal vein

Vestibulocochlear ganglion (VIII)

Otic vesicle
Epithelium

Lumen

Petrous temporal bone

*Cell sparse
superarachnoid
reticulum*

Nerve IX
(glosso-
pharyngeal)

Nerve and
ganglion X
(vagal)

Section 172 brain *in situ*

0.5 mm

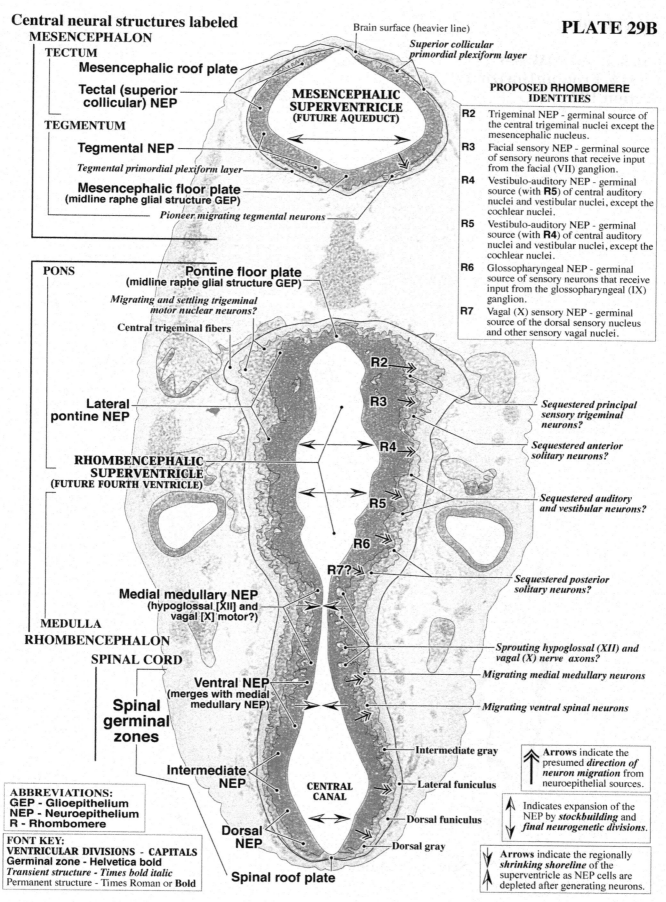

79

PLATE 29B

Central neural structures labeled

MESENCEPHALON
TECTUM
Mesencephalic roof plate
Tectal (superior collicular) NEP
TEGMENTUM
Tegmental NEP
Tegmental primordial plexiform layer
Mesencephalic floor plate
(midline raphe glial structure GEP)
Pioneer migrating tegmental neurons

Brain surface (heavier line)
Superior collicular primordial plexiform layer

MESENCEPHALIC SUPERVENTRICLE
(FUTURE AQUEDUCT)

PROPOSED RHOMBOMERE IDENTITIES

R2 Trigeminal NEP - germinal source of the central trigeminal nuclei except the mesencephalic nucleus.

R3 Facial sensory NEP - germinal source of sensory neurons that receive input from the facial (VII) ganglion.

R4 Vestibulo-auditory NEP - germinal source (with **R5**) of central auditory nuclei and vestibular nuclei, except the cochlear nuclei.

R5 Vestibulo-auditory NEP - germinal source (with **R4**) of central auditory nuclei and vestibular nuclei, except the cochlear nuclei.

R6 Glossopharyngeal NEP - germinal source of sensory neurons that receive input from the glossopharyngeal (IX) ganglion.

R7 Vagal (X) sensory NEP - germinal source of the dorsal sensory nucleus and other sensory vagal nuclei.

PONS
Pontine floor plate
(midline raphe glial structure GEP)
Migrating and settling trigeminal motor nuclear neurons?
Central trigeminal fibers

Lateral pontine NEP

RHOMBENCEPHALIC SUPERVENTRICLE
(FUTURE FOURTH VENTRICLE)

R2
R3
R4
R5
R6
R7?

Sequestered principal sensory trigeminal neurons?
Sequestered anterior solitary neurons?
Sequestered auditory and vestibular neurons?
Sequestered posterior solitary neurons?

Medial medullary NEP
(hypoglossal [XII] and vagal [X] motor?)

MEDULLA
RHOMBENCEPHALON
SPINAL CORD

Spinal germinal zones

Ventral NEP
(merges with medial medullary NEP)

Sprouting hypoglossal (XII) and vagal (X) nerve axons?
Migrating medial medullary neurons
Migrating ventral spinal neurons

Intermediate gray

Intermediate NEP

CENTRAL CANAL

Lateral funiculus

Dorsal funiculus

Dorsal NEP

Dorsal gray

Spinal roof plate

ABBREVIATIONS:
GEP - Glioepithelium
NEP - Neuroepithelium
R - Rhombomere

FONT KEY:
VENTRICULAR DIVISIONS - CAPITALS
Germinal zone - Helvetica bold
Transient structure - Times bold italic
Permanent structure - Times Roman or **Bold**

Arrows indicate the presumed *direction of neuron migration* from neuroepithelial sources.

Indicates expansion of the NEP by *stockbuilding* and *final neurogenetic divisions*.

Arrows indicate the regionally *shrinking shoreline* of the superventricle as NEP cells are depleted after generating neurons.

PLATE 30A

CR 8.0 mm, GW 6.0
C8314, Frontal/Horizontal
Section 182

**Peripheral neural
and non-neural
structures labeled**

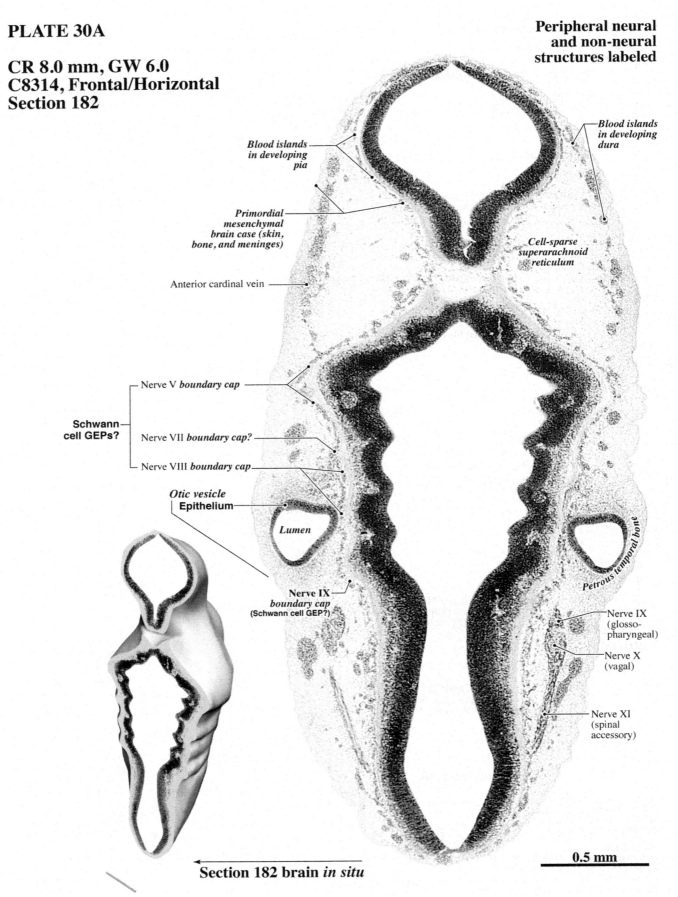

*Blood islands
in developing
pia*

*Blood islands
in developing
dura*

*Primordial
mesenchymal
brain case (skin,
bone, and meninges)*

*Cell-sparse
superarachnoid
reticulum*

Anterior cardinal vein

Nerve V *boundary cap*

**Schwann
cell GEPs?**

Nerve VII *boundary cap?*

Nerve VIII *boundary cap*

Otic vesicle
Epithelium

Lumen

Nerve IX
boundary cap
(Schwann cell GEP?)

Petrous temporal bone

Nerve IX
(glosso-
pharyngeal)

Nerve X
(vagal)

Nerve XI
(spinal
accessory)

Section 182 brain *in situ*

0.5 mm

Central neural structures labeled

PLATE 30B

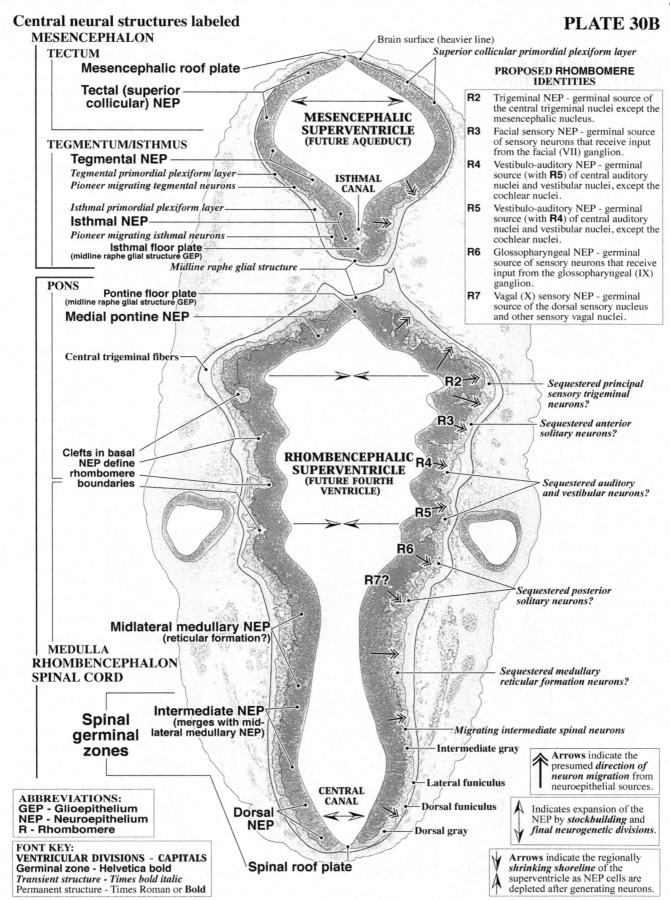

MESENCEPHALON

TECTUM

Mesencephalic roof plate

Tectal (superior collicular) NEP

Brain surface (heavier line)

Superior collicular primordial plexiform layer

MESENCEPHALIC SUPERVENTRICLE (FUTURE AQUEDUCT)

ISTHMAL CANAL

TEGMENTUM/ISTHMUS

Tegmental NEP

Tegmental primordial plexiform layer

Pioneer migrating tegmental neurons

Isthmal primordial plexiform layer

Isthmal NEP

Pioneer migrating isthmal neurons

Isthmal floor plate
(midline raphe glial structure GEP)

Midline raphe glial structure

PONS

Pontine floor plate
(midline raphe glial structure GEP)

Medial pontine NEP

Central trigeminal fibers

RHOMBENCEPHALIC SUPERVENTRICLE (FUTURE FOURTH VENTRICLE)

Clefts in basal NEP define rhombomere boundaries

R2

R3

R4

R5

R6

R7?

Sequestered principal sensory trigeminal neurons?

Sequestered anterior solitary neurons?

Sequestered auditory and vestibular neurons?

Sequestered posterior solitary neurons?

Midlateral medullary NEP
(reticular formation?)

Sequestered medullary reticular formation neurons?

MEDULLA
RHOMBENCEPHALON
SPINAL CORD

Spinal germinal zones

Intermediate NEP
(merges with mid-lateral medullary NEP)

Migrating intermediate spinal neurons

Intermediate gray

Lateral funiculus

Dorsal funiculus

Dorsal gray

Dorsal NEP

CENTRAL CANAL

Spinal roof plate

PROPOSED RHOMBOMERE IDENTITIES

R2	Trigeminal NEP - germinal source of the central trigeminal nuclei except the mesencephalic nucleus.
R3	Facial sensory NEP - germinal source of sensory neurons that receive input from the facial (VII) ganglion.
R4	Vestibulo-auditory NEP - germinal source (with **R5**) of central auditory nuclei and vestibular nuclei, except the cochlear nuclei.
R5	Vestibulo-auditory NEP - germinal source (with **R4**) of central auditory nuclei and vestibular nuclei, except the cochlear nuclei.
R6	Glossopharyngeal NEP - germinal source of sensory neurons that receive input from the glossopharyngeal (IX) ganglion.
R7	Vagal (X) sensory NEP - germinal source of the dorsal sensory nucleus and other sensory vagal nuclei.

ABBREVIATIONS:
GEP - Glioepithelium
NEP - Neuroepithelium
R - Rhombomere

FONT KEY:
VENTRICULAR DIVISIONS - CAPITALS
Germinal zone - Helvetica bold
Transient structure - Times bold italic
Permanent structure - Times Roman or **Bold**

Arrows indicate the presumed *direction of neuron migration* from neuroepithelial sources.

Indicates expansion of the NEP by *stockbuilding* and *final neurogenetic divisions*.

Arrows indicate the regionally *shrinking shoreline* of the superventricle as NEP cells are depleted after generating neurons.

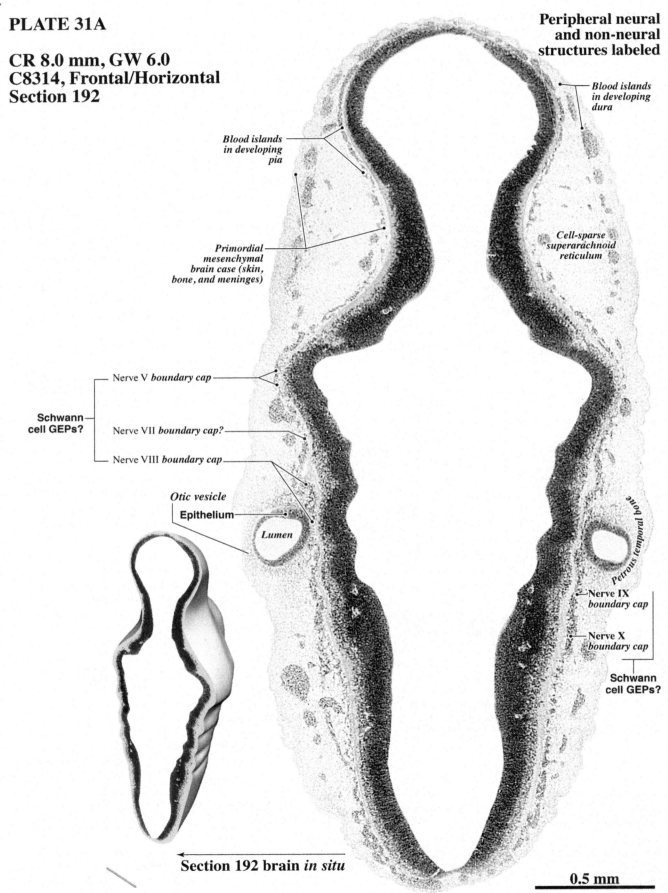

PLATE 31A

**CR 8.0 mm, GW 6.0
C8314, Frontal/Horizontal
Section 192**

**Peripheral neural
and non-neural
structures labeled**

*Blood islands
in developing
dura*

*Blood islands
in developing
pia*

*Cell-sparse
superarachnoid
reticulum*

*Primordial
mesenchymal
brain case (skin,
bone, and meninges)*

Nerve V *boundary cap*

**Schwann
cell GEPs?**

Nerve VII *boundary cap?*

Nerve VIII *boundary cap*

Otic vesicle

Epithelium

Lumen

Petrous temporal bone

Nerve IX
boundary cap

Nerve X
boundary cap

**Schwann
cell GEPs?**

Section 192 brain *in situ*

0.5 mm

Central neural structures labeled

PLATE 31B

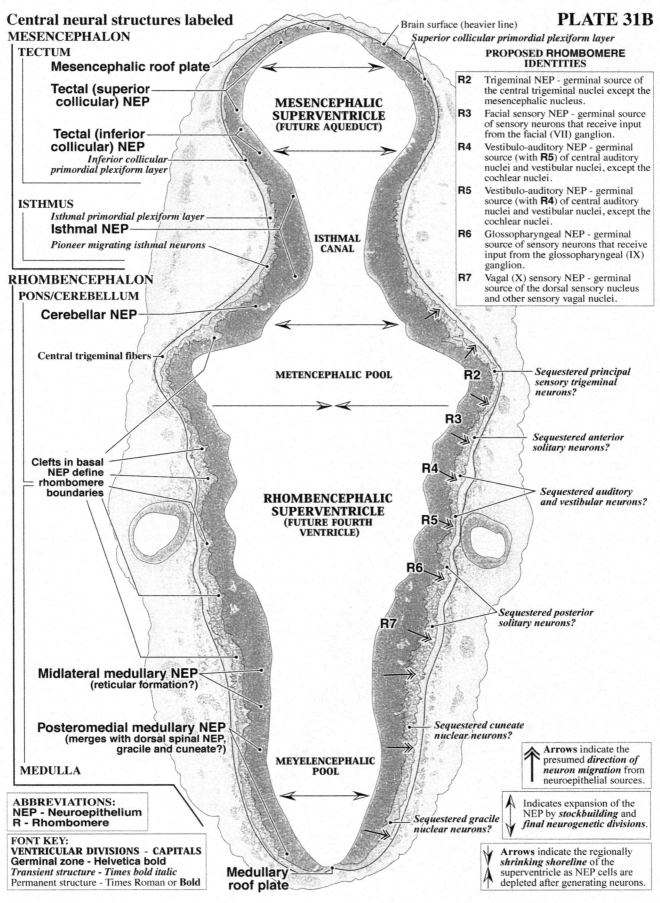

MESENCEPHALON

TECTUM

Mesencephalic roof plate

Tectal (superior collicular) NEP

Tectal (inferior collicular) NEP

Inferior collicular primordial plexiform layer

ISTHMUS

Isthmal primordial plexiform layer

Isthmal NEP

Pioneer migrating isthmal neurons

RHOMBENCEPHALON

PONS/CEREBELLUM

Cerebellar NEP

Central trigeminal fibers

Clefts in basal NEP define rhombomere boundaries

Midlateral medullary NEP (reticular formation?)

Posteromedial medullary NEP (merges with dorsal spinal NEP, gracile and cuneate?)

MEDULLA

Brain surface (heavier line)

Superior collicular primordial plexiform layer

MESENCEPHALIC SUPERVENTRICLE (FUTURE AQUEDUCT)

ISTHMAL CANAL

METENCEPHALIC POOL

RHOMBENCEPHALIC SUPERVENTRICLE (FUTURE FOURTH VENTRICLE)

MEYELENCEPHALIC POOL

Medullary roof plate

PROPOSED RHOMBOMERE IDENTITIES

R2 Trigeminal NEP - germinal source of the central trigeminal nuclei except the mesencephalic nucleus.

R3 Facial sensory NEP - germinal source of sensory neurons that receive input from the facial (VII) ganglion.

R4 Vestibulo-auditory NEP - germinal source (with **R5**) of central auditory nuclei and vestibular nuclei, except the cochlear nuclei.

R5 Vestibulo-auditory NEP - germinal source (with **R4**) of central auditory nuclei and vestibular nuclei, except the cochlear nuclei.

R6 Glossopharyngeal NEP - germinal source of sensory neurons that receive input from the glossopharyngeal (IX) ganglion.

R7 Vagal (X) sensory NEP - germinal source of the dorsal sensory nucleus and other sensory vagal nuclei.

R2 *Sequestered principal sensory trigeminal neurons?*

R3 *Sequestered anterior solitary neurons?*

R4 *Sequestered auditory and vestibular neurons?*

R5

R6 *Sequestered posterior solitary neurons?*

R7

Sequestered cuneate nuclear neurons?

Sequestered gracile nuclear neurons?

ABBREVIATIONS:
NEP - Neuroepithelium
R - Rhombomere

FONT KEY:
VENTRICULAR DIVISIONS - CAPITALS
Germinal zone - Helvetica bold
Transient structure - Times bold italic
Permanent structure - Times Roman or **Bold**

Arrows indicate the presumed *direction of neuron migration* from neuroepithelial sources.

Indicates expansion of the NEP by *stockbuilding* and *final neurogenetic divisions*.

Arrows indicate the regionally *shrinking shoreline* of the superventricle as NEP cells are depleted after generating neurons.

84

PLATE 32A

CR 8.0 mm, GW 6.0
C8314, Frontal/Horizontal
Section 222

Peripheral neural
and non-neural
structures labeled

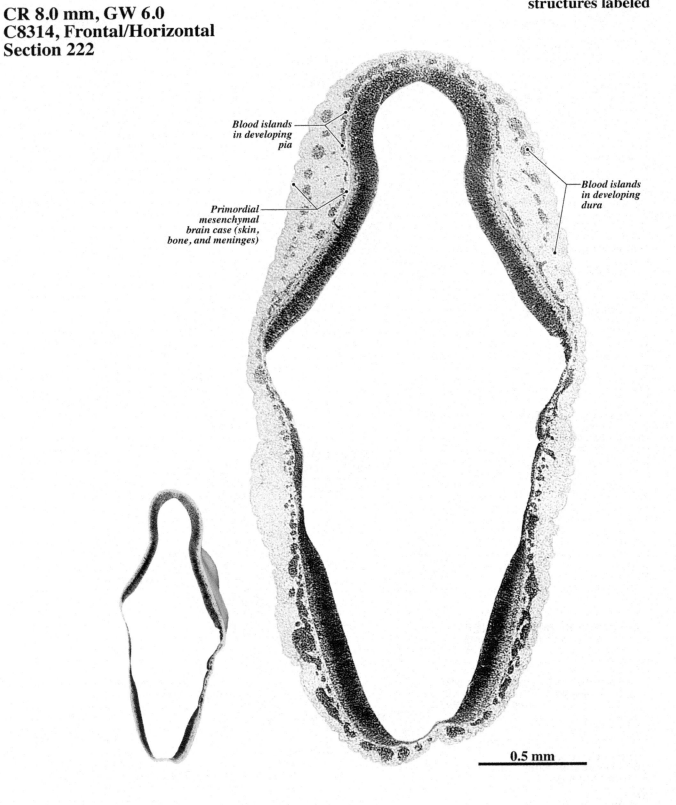

Blood islands in developing pia

Primordial mesenchymal brain case (skin, bone, and meninges)

Blood islands in developing dura

0.5 mm

← **Section 222 brain** *in situ*

Central neural structures labeled

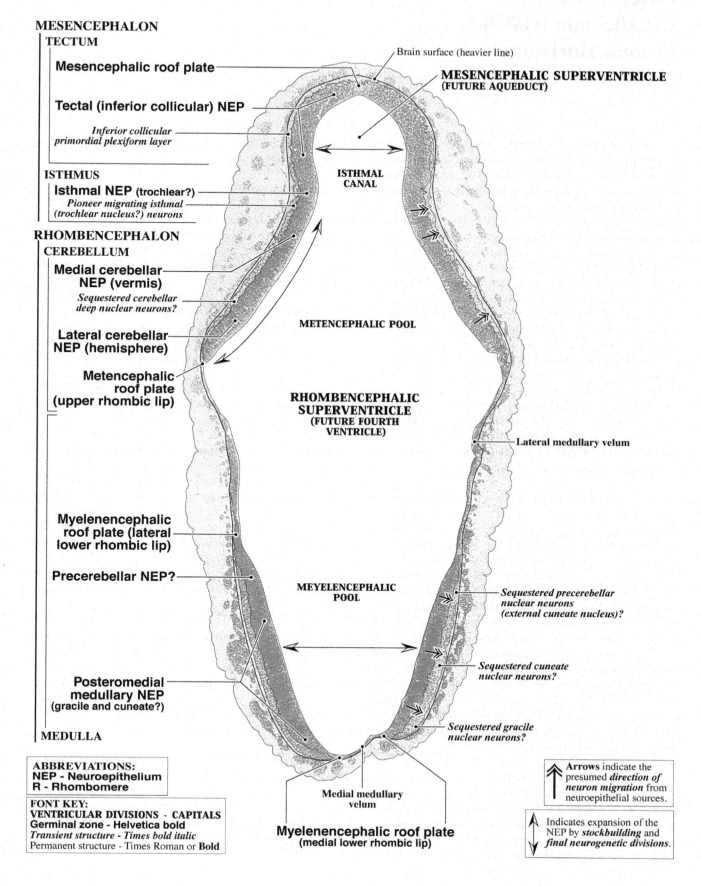

MESENCEPHALON

TECTUM

Mesencephalic roof plate

Tectal (inferior collicular) NEP

Inferior collicular primordial plexiform layer

ISTHMUS

Isthmal NEP (trochlear?)

Pioneer migrating isthmal (trochlear nucleus?) neurons

RHOMBENCEPHALON

CEREBELLUM

Medial cerebellar NEP (vermis)

Sequestered cerebellar deep nuclear neurons?

Lateral cerebellar NEP (hemisphere)

Metencephalic roof plate (upper rhombic lip)

Myelenencephalic roof plate (lateral lower rhombic lip)

Precerebellar NEP?

Posteromedial medullary NEP (gracile and cuneate?)

MEDULLA

Brain surface (heavier line)

MESENCEPHALIC SUPERVENTRICLE (FUTURE AQUEDUCT)

ISTHMAL CANAL

METENCEPHALIC POOL

RHOMBENCEPHALIC SUPERVENTRICLE (FUTURE FOURTH VENTRICLE)

Lateral medullary velum

MEYELENCEPHALIC POOL

Sequestered precerebellar nuclear neurons (external cuneate nucleus)?

Sequestered cuneate nuclear neurons?

Sequestered gracile nuclear neurons?

Medial medullary velum

Myelenencephalic roof plate (medial lower rhombic lip)

ABBREVIATIONS:
NEP - Neuroepithelium
R - Rhombomere

FONT KEY:
VENTRICULAR DIVISIONS - CAPITALS
Germinal zone - Helvetica bold
Transient structure - Times bold italic
Permanent structure - Times Roman or **Bold**

Arrows indicate the presumed *direction of neuron migration* from neuroepithelial sources.

Indicates expansion of the NEP by *stockbuilding* and *final neurogenetic divisions*.

PART V: C6516
CR 10.5 mm (GW 6.5)
Frontal/Horizontal

Carnegie Collection specimen #6516 (designated here as C6516) with a 10.5-mm crown-rump length (CR) is estimated to be at gestational week (GW) 6.5. C6516 was fixed in corrosive acetic acid, embedded in a celloidin/paraffin mix, and was cut in 8-μm sagittal sections that were stained with aluminum cochineal. Various orientations of the computer-aided 3-D reconstruction of M1000's brain are used to show the gross external features of a GW5.5 brain (**Figure 13**). Like most sagittally cut specimens, C6516's sections are not parallel to the midline; **Figure 16** shows the approximate rotations in front (**B**) and back views (**C**). We photographed 65 sections at low magnification from the left to right sides of the brain. Five of the sections, mainly from the left side of the brain, are illustrated in **Plates 33AB to 37AB**. Each illustrated section shows the brain with all surrounding tissues. Labels in **A Plates** (normal-contrast images) identify the approximate midline, non-neural structures, peripheral neural structures, and brain ventricular divisions; labels in **B Plates** (low-contrast images) identify central neural structures. **Plates 38AB to 47AB** show high-magnification views of many parts of the developing brain.

The telencephalon is the smallest major brain structure surrounding an expanding telencephalic superventricle. The cortical neuroepithelium (NEP) is still in stockbuilding stage, but the other NEPs are well into the neurogenetic phase. In the cortex, a primordial plexiform layer consists of discontinuous cell-sparse areas with some pioneer settling Cajal-Retzius neurons. The basal cortical NEP is postulated to have postmitotic Cajal-Retzius and subplate neurons sequestered there prior to migration (**Plate 38**). In contrast, the slightly younger next specimen (M1000) has very few cells in the primordial plexiform layer. Some migrating neurons are adjacent to the basal ganglionic and basal telencephalic NEPs and are beginning to form mounds in the floor of the telencephalon. The olfactory epithelium is invaginating into the nasal cavity and tentatively identified olfactory nerve fibers are growing toward the brain amid some dense mesenchymal cells.

The diencephalon is the larger forebrain structure. The thalamic NEP is stockbuilding as it surrounds a dorsally expanding superventricle. The hypothalamic and subthalamic NEPs areas are depleting their population of stem cells. Postmitotic, premigratory younger neurons are postulated to be sequestered in basal parts of the NEP, but the

many neurons that have already been generated are accumulating in the parenchyma.

The mesencephalon forms a prominent arch around the mesencephalic flexure. The roof (tectum and pretectum) of the mesencephalon contains a stockbuilding neuroepithelium adjacent to a thin cell-sparse layer; the population of stem cells is rapidly adding new members and the tectum grows over the ballooning mesencephalic superventricle. The tegmental and isthmal NEPs are rapidly unloading their neuronal progeny that settle in dense bands in the adjacent parenchyma. Many neurons that will reside here in the mature brain have already been generated, but they have not settled into any recognizable nuclear structures (the oculomotor nuclei, for instance).

The rhombencephalon is the largest brain structure. Both the pons and medulla have neuroepithelia that form crescent-shaped rhombomeres in lateral areas. In the sagittal plane, it is easy to see that rhombomeres are unloading their neuronal and glial progeny into parenchymal expansions at the entry zones of sensory cranial nerves V, VII, VIII, IX, and X. Neurons migrating in these areas are tentatively identified as receptors of the incoming sensory axons. For example, trigeminal nuclear neurons (mainly those in the principal sensory nucleus) are generated in rhombomere 2 and migrate outward to mingle with incoming afferents from the trigeminal ganglion. Medially, the pons and medulla contain longitudinal bands of migrating cells, but nuclear subdivisions are generally absent in the parenchyma. The genu of the facial motor nerve forms fascicles adjacent to the rhombomere 3 NEP (**Plate 35B**). Rhombomere 3 is the putative source of neurons that will be receptive to axons of the facial ganglion; eventually, these neurons will settle in the anterior solitary nucleus. The subpial fiber band is definitely thicker in lateral areas where the axons from sensory ganglia enter the brain. Peripheral nerves have dense glia (Schwann cells), while central fiber tracts are clear.

The cerebellum stands out as the most immature rhombencephalic structure. All parts of the cerebellar NEP are stockbuilding neuronal and glial stem cells. But some progenitors in the cerebellar NEP are in neurogenetic phase, producing deep nuclear neurons that occupy distinct layers in the cerebellar transitional field.

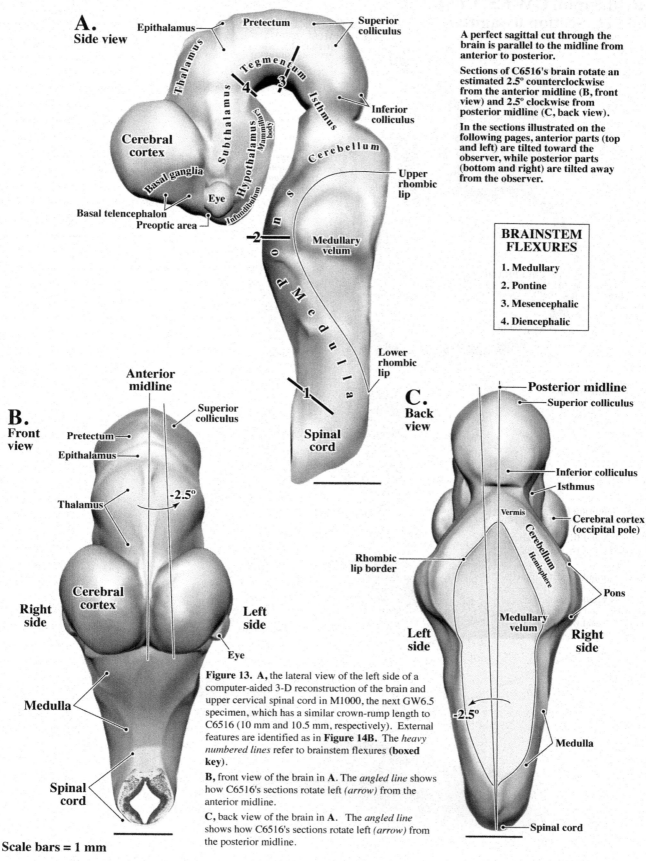

EXTERNAL FEATURES OF THE GW6.5 BRAIN

A.
Side view

Epithalamus Pretectum Superior colliculus

Thalamus

Tegmentum

Isthmus

Subthalamus

Hypothalamus

Mammillary body

Cerebral cortex

Basal ganglia

Basal telencephalon

Preoptic area

Eye

Infundibulum

Cerebellum

Inferior colliculus

Upper rhombic lip

Pons

Medulla

Medullary velum

Lower rhombic lip

Spinal cord

A perfect sagittal cut through the brain is parallel to the midline from anterior to posterior.

Sections of C6516's brain rotate an estimated 2.5° counterclockwise from the anterior midline (B, front view) and 2.5° clockwise from posterior midline (C, back view).

In the sections illustrated on the following pages, anterior parts (top and left) are tilted toward the observer, while posterior parts (bottom and right) are tilted away from the observer.

BRAINSTEM FLEXURES
1. Medullary
2. Pontine
3. Mesencephalic
4. Diencephalic

B.
Front view

Anterior midline

Superior colliculus

Pretectum

Epithalamus

Thalamus

-2.5°

Cerebral cortex

Right side

Left side

Eye

Medulla

Spinal cord

C.
Back view

Posterior midline

Superior colliculus

Inferior colliculus

Isthmus

Vermis

Cerebral cortex (occipital pole)

Cerebellum Hemisphere

Rhombic lip border

Pons

Medullary velum

Left side

Right side

-2.5°

Medulla

Spinal cord

Figure 13. A, the lateral view of the left side of a computer-aided 3-D reconstruction of the brain and upper cervical spinal cord in M1000, the next GW6.5 specimen, which has a similar crown-rump length to C6516 (10 mm and 10.5 mm, respectively). External features are identified as in **Figure 14B.** The *heavy numbered lines* refer to brainstem flexures **(boxed key).**

B, front view of the brain in **A.** The *angled line* shows how C6516's sections rotate left *(arrow)* from the anterior midline.

C, back view of the brain in **A.** The *angled line* shows how C6516's sections rotate left *(arrow)* from the posterior midline.

Scale bars = 1 mm

88

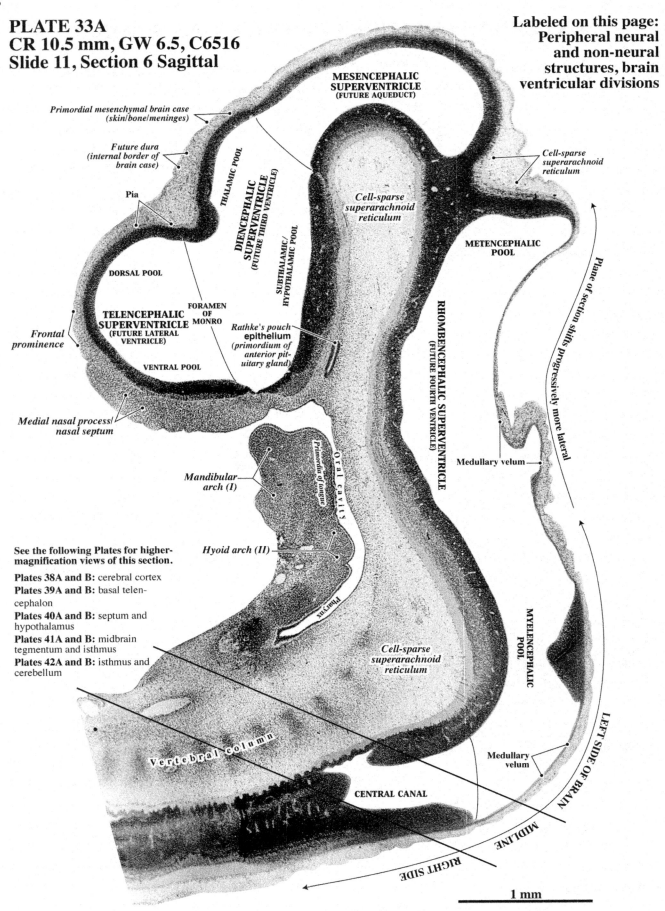

PLATE 33A
CR 10.5 mm, GW 6.5, C6516
Slide 11, Section 6 Sagittal

Labeled on this page:
Peripheral neural
and non-neural
structures, brain
ventricular divisions

Primordial mesenchymal brain case
(skin/bone/meninges)

Future dura
(internal border of
brain case)

Pia

DORSAL POOL

Frontal
prominence

TELENCEPHALIC
SUPERVENTRICLE
(FUTURE LATERAL
VENTRICLE)

FORAMEN
OF
MONRO

VENTRAL POOL

Medial nasal process/
nasal septum

THALAMIC POOL

DIENCEPHALIC
SUPERVENTRICLE
(FUTURE THIRD VENTRICLE)

SUBTHALAMIC/
HYPOTHALAMIC POOL

Rathke's pouch epithelium
(primordium of
anterior pit-
uitary gland)

MESENCEPHALIC
SUPERVENTRICLE
(FUTURE AQUEDUCT)

Cell-sparse
superarachnoid
reticulum

Cell-sparse
superarachnoid
reticulum

METENCEPHALIC
POOL

RHOMBENCEPHALIC SUPERVENTRICLE
(FUTURE FOURTH VENTRICLE)

Plane of section shifts progressively more lateral

Medullary velum

Mandibular
arch (I)

Primordia of tongue

Oral cavity

Hyoid arch (II)

Pharynx

See the following Plates for higher-
magnification views of this section.

Plates 38A and B: cerebral cortex
Plates 39A and B: basal telencephalon
Plates 40A and B: septum and
hypothalamus
Plates 41A and B: midbrain
tegmentum and isthmus
Plates 42A and B: isthmus and
cerebellum

Cell-sparse
superarachnoid
reticulum

MYELENCEPHALIC
POOL

Medullary
velum

Vertebral column

CENTRAL CANAL

LEFT SIDE OF BRAIN

MIDLINE

RIGHT SIDE

1 mm

Labeled on this page:
Central neural structures

The cortical, thalamic, tectal, and cerebellar NEPs form *expanding shorelines* of the superventricle as stockbuilding NEP cells increase.

The basal telencephalic, septal, preoptic, hypothalamic, subthalamic, tegmental, pontine, and medullary NEPs are *shrinking* as they produce neurons.

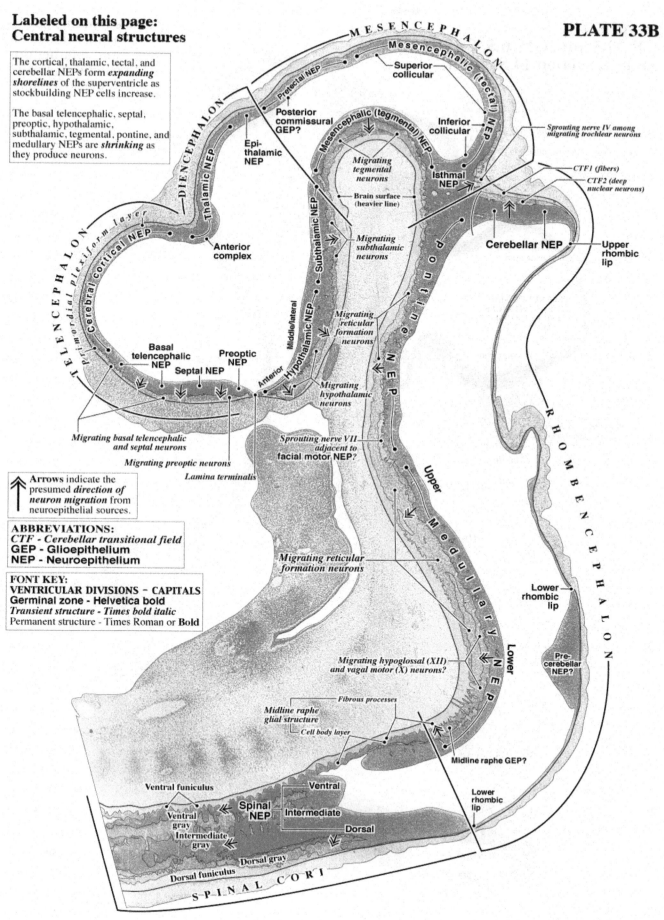

MESENCEPHALON

Mesencephalic (tectal) NEP

Superior collicular

Pretectal NEP

Posterior commissural GEP?

Inferior collicular

Mesencephalic (tegmental) NEP

Sprouting nerve IV among migrating trochlear neurons

DIENCEPHALON

Epi-thalamic NEP

Thalamic NEP

Migrating tegmental neurons

Isthmal NEP

CTF1 (fibers)

CTF2 (deep nuclear neurons)

Brain surface (heavier line)

Cerebellar NEP

Upper rhombic lip

TELENCEPHALON

Primordial plexiform layer

Cerebral cortical NEP

Anterior complex

Subthalamic NEP

Migrating subthalamic neurons

Pontine NEP

Middle/lateral

Hypothalamic NEP

Migrating reticular formation neurons

Basal telencephalic NEP

Preoptic NEP

Septal NEP

Anterior

Migrating hypothalamic neurons

Upper

RHOMBENCEPHALON

Migrating basal telencephalic and septal neurons

Migrating preoptic neurons

Lamina terminalis

Sprouting nerve VII adjacent to facial motor NEP?

Medullary NEP

↑ Arrows indicate the presumed *direction of neuron migration* from neuroepithelial sources.

ABBREVIATIONS:
CTF - Cerebellar transitional field
GEP - Glioepithelium
NEP - Neuroepithelium

Migrating reticular formation neurons

Lower rhombic lip

Pre-cerebellar NEP?

FONT KEY:
VENTRICULAR DIVISIONS – CAPITALS
Germinal zone - Helvetica bold
Transient structure - Times bold italic
Permanent structure - Times Roman or **Bold**

Migrating hypoglossal (XII) and vagal motor (X) neurons?

Lower

Fibrous processes

Midline raphe glial structure

Cell body layer

Midline raphe GEP?

Lower rhombic lip

Ventral funiculus

Ventral

Spinal NEP

Ventral gray

Intermediate

Intermediate gray

Dorsal

Dorsal gray

Dorsal funiculus

SPINAL CORD

PLATE 34A
CR 10.5 mm, GW 6.5, C6516
Slide 9, Section 14 Sagittal

Labeled on this page:
Peripheral neural
and non-neural
structures, brain
ventricular divisions

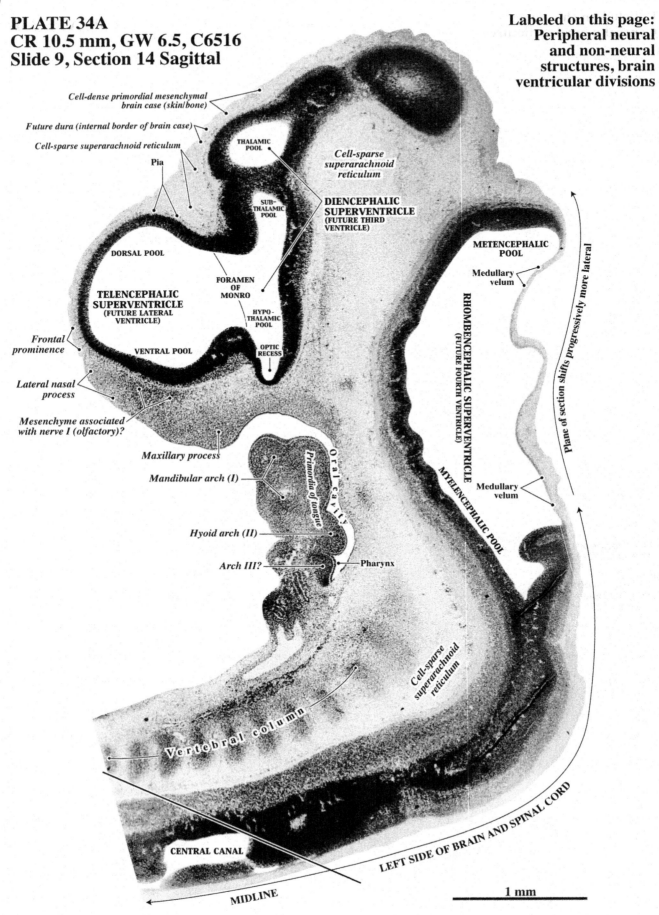

Cell-dense primordial mesenchymal
brain case (skin/bone)

Future dura (internal border of brain case)

Cell-sparse superarachnoid reticulum

Pia

THALAMIC
POOL

Cell-sparse
superarachnoid
reticulum

SUB-
THALAMIC
POOL

**DIENCEPHALIC
SUPERVENTRICLE
(FUTURE THIRD
VENTRICLE)**

DORSAL POOL

METENCEPHALIC
POOL

Medullary
velum

**TELENCEPHALIC
SUPERVENTRICLE
(FUTURE LATERAL
VENTRICLE)**

FORAMEN
OF
MONRO

HYPO-
THALAMIC
POOL

Frontal
prominence

VENTRAL POOL

OPTIC
RECESS

**RHOMBENCEPHALIC SUPERVENTRICLE
(FUTURE FOURTH VENTRICLE)**

Lateral nasal
process

Mesenchyme associated
with nerve I (olfactory)?

Maxillary process

Primordia of tongue

Oral cavity

Mandibular arch (I)

Hyoid arch (II)

Arch III?

Pharynx

MYELENCEPHALIC POOL

Medullary
velum

Plane of section shifts progressively more lateral

Cell-sparse
superarachnoid
reticulum

Vertebral column

CENTRAL CANAL

LEFT SIDE OF BRAIN AND SPINAL CORD

MIDLINE

1 mm

Labeled on this page:
Central neural structures

The cortical, thalamic, tectal, and cerebellar NEPs form *expanding shorelines* of the superventricles as stockbuilding NEP cells increase.

The basal telencephalic, hypothalamic, subthalamic, pontine, and medullary NEPs *shrink* as cells in neurogenetic phase produce neurons.

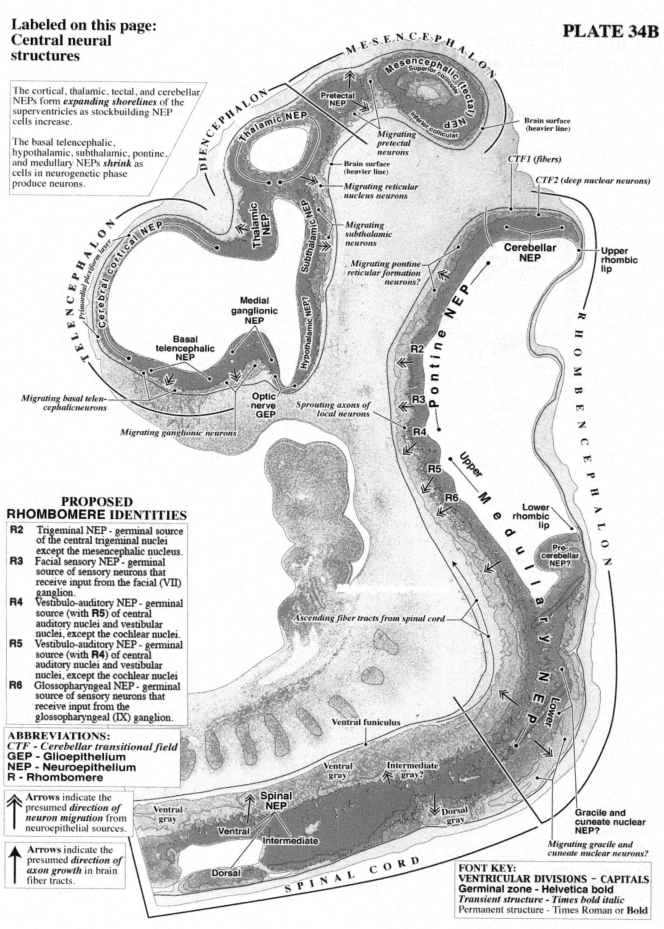

MESENCEPHALON

Mesencephalic (tectal) NEP
Superior colliculus
Inferior collicular

Pretectal NEP

DIENCEPHALON

Thalamic NEP

Migrating pretectal neurons

Brain surface (heavier line)

CTF1 (fibers)

CTF2 (deep nuclear neurons)

Thalamic NEP

Brain surface (heavier line)

Migrating reticular nucleus neurons

Cerebellar NEP

Upper rhombic lip

TELENCEPHALON

Cerebral cortical NEP
Primordial plexiform layer

Subthalamic NEP

Migrating subthalamic neurons

Migrating pontine reticular formation neurons?

Pontine NEP

RHOMBENCEPHALON

Medial ganglionic NEP

Hypothalamic NEP?

R2

Basal telencephalic NEP

R3

Upper Medulla ry NEP

Optic nerve GEP

Sprouting axons of local neurons

R4

Migrating basal telen-cephalic neurons

R5

R6

Lower rhombic lip

Migrating ganglionic neurons

Pre-cerebellar NEP?

PROPOSED
RHOMBOMERE IDENTITIES

R2 Trigeminal NEP - germinal source of the central trigeminal nuclei except the mesencephalic nucleus.

R3 Facial sensory NEP - germinal source of sensory neurons that receive input from the facial (VII) ganglion.

R4 Vestibulo-auditory NEP - germinal source (with **R5**) of central auditory nuclei and vestibular nuclei, except the cochlear nuclei.

R5 Vestibulo-auditory NEP - germinal source (with **R4**) of central auditory nuclei and vestibular nuclei, except the cochlear nuclei

R6 Glossopharyngeal NEP - germinal source of sensory neurons that receive input from the glossopharyngeal (IX) ganglion.

Ascending fiber tracts from spinal cord

Lower Medulla ry NEP

Gracile and cuneate nuclear NEP?

Migrating gracile and cuneate nuclear neurons?

ABBREVIATIONS:
CTF - Cerebellar transitional field
GEP - Glioepithelium
NEP - Neuroepithelium
R - Rhombomere

Ventral funiculus

Intermediate gray?

Dorsal gray

Arrows indicate the presumed *direction of neuron migration* from neuroepithelial sources.

Ventral gray

Spinal NEP

Ventral gray

Arrows indicate the presumed *direction of axon growth* in brain fiber tracts.

Ventral

Intermediate

Dorsal

SPINAL CORD

FONT KEY:
VENTRICULAR DIVISIONS – CAPITALS
Germinal zone - Helvetica bold
Transient structure - Times bold italic
Permanent structure - Times Roman or **Bold**

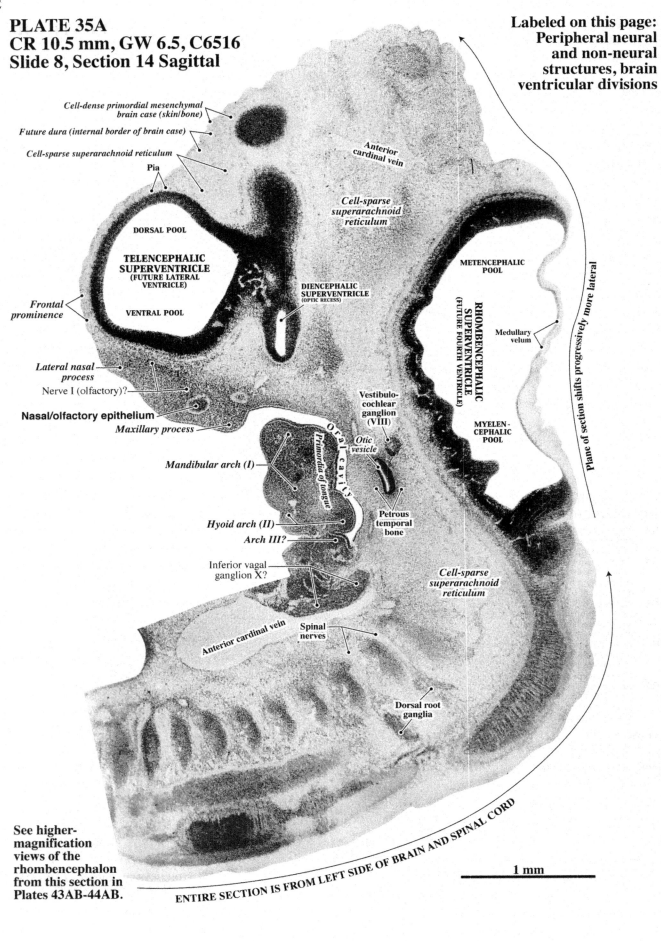

92

PLATE 35A
CR 10.5 mm, GW 6.5, C6516
Slide 8, Section 14 Sagittal

Labeled on this page:
Peripheral neural
and non-neural
structures, brain
ventricular divisions

Cell-dense primordial mesenchymal
brain case (skin/bone)

Future dura (internal border of brain case)

Cell-sparse superarachnoid reticulum

Pia

Anterior
cardinal vein

Cell-sparse
superarachnoid
reticulum

DORSAL POOL

TELENCEPHALIC
SUPERVENTRICLE
(FUTURE LATERAL
VENTRICLE)

METENCEPHALIC
POOL

DIENCEPHALIC
SUPERVENTRICLE
(OPTIC RECESS)

Frontal
prominence

VENTRAL POOL

RHOMBENCEPHALIC
SUPERVENTRICLE
(FUTURE FOURTH VENTRICLE)

Medullary
velum

Lateral nasal
process

Nerve I (olfactory)?

Nasal/olfactory epithelium

Vestibulo-
cochlear
ganglion
(VIII)

MYELEN-
CEPHALIC
POOL

Otic
vesicle

Maxillary process

Oral cavity

Primordia of tongue

Mandibular arch (I)

Hyoid arch (II)

Petrous
temporal
bone

Arch III?

Inferior vagal
ganglion X?

Cell-sparse
superarachnoid
reticulum

Spinal
nerves

Anterior cardinal vein

Dorsal root
ganglia

Plane of section shifts progressively more lateral

See higher-
magnification
views of the
rhombencephalon
from this section in
Plates 43AB-44AB.

ENTIRE SECTION IS FROM LEFT SIDE OF BRAIN AND SPINAL CORD

1 mm

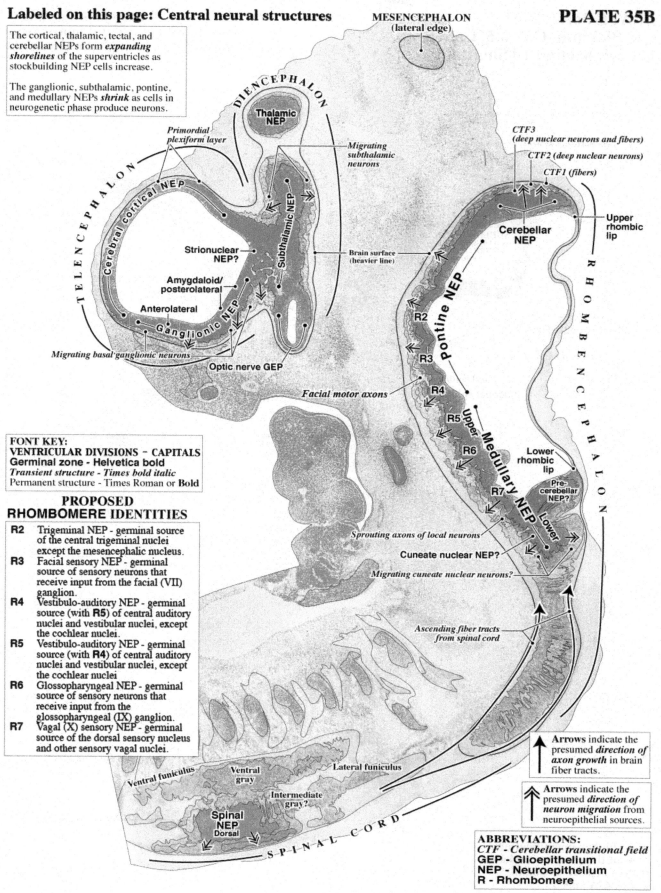

Labeled on this page: Central neural structures

The cortical, thalamic, tectal, and cerebellar NEPs form *expanding shorelines* of the superventricles as stockbuilding NEP cells increase.

The ganglionic, subthalamic, pontine, and medullary NEPs *shrink* as cells in neurogenetic phase produce neurons.

MESENCEPHALON (lateral edge)

DIENCEPHALON

Thalamic NEP

Primordial plexiform layer

Migrating subthalamic neurons

CTF3 *(deep nuclear neurons and fibers)*

CTF2 *(deep nuclear neurons)*

CTF1 *(fibers)*

TELENCEPHALON

Cerebral cortical NEP

Strionuclear NEP?

Subthalamic NEP

Amygdaloid/ posterolateral

Anterolateral

Ganglionic NEP

Migrating basal ganglionic neurons

Optic nerve GEP

Brain surface (heavier line)

Cerebellar NEP

Upper rhombic lip

RHOMBENCEPHALON

R2

Pontine NEP

R3

R4

Facial motor axons

R5

Upper

R6

Medullary NEP

Lower rhombic lip

Pre-cerebellar NEP?

R7

Lower

Sprouting axons of local neurons

Cuneate nuclear NEP?

Migrating cuneate nuclear neurons?

Ascending fiber tracts from spinal cord

FONT KEY:
VENTRICULAR DIVISIONS – CAPITALS
Germinal zone - Helvetica bold
Transient structure - Times bold italic
Permanent structure - Times Roman or **Bold**

PROPOSED RHOMBOMERE IDENTITIES

R2 Trigeminal NEP - germinal source of the central trigeminal nuclei except the mesencephalic nucleus.

R3 Facial sensory NEP - germinal source of sensory neurons that receive input from the facial (VII) ganglion.

R4 Vestibulo-auditory NEP - germinal source (with **R5**) of central auditory nuclei and vestibular nuclei, except the cochlear nuclei.

R5 Vestibulo-auditory NEP - germinal source (with **R4**) of central auditory nuclei and vestibular nuclei, except the cochlear nuclei.

R6 Glossopharyngeal NEP - germinal source of sensory neurons that receive input from the glossopharyngeal (IX) ganglion.

R7 Vagal (X) sensory NEP - germinal source of the dorsal sensory nucleus and other sensory vagal nuclei.

Ventral funiculus

Ventral gray

Intermediate gray?

Lateral funiculus

Spinal NEP Dorsal

SPINAL CORD

Arrows indicate the presumed *direction of axon growth* in brain fiber tracts.

Arrows indicate the presumed *direction of neuron migration* from neuroepithelial sources.

ABBREVIATIONS:
CTF - Cerebellar transitional field
GEP - Glioepithelium
NEP - Neuroepithelium
R - Rhombomere

PLATE 36A
CR 10.5 mm, GW 6.5, C6516
Slide 7, Section 10 Sagittal

Labeled on this page:
Peripheral neural and non-neural
structures, brain ventricular divisions

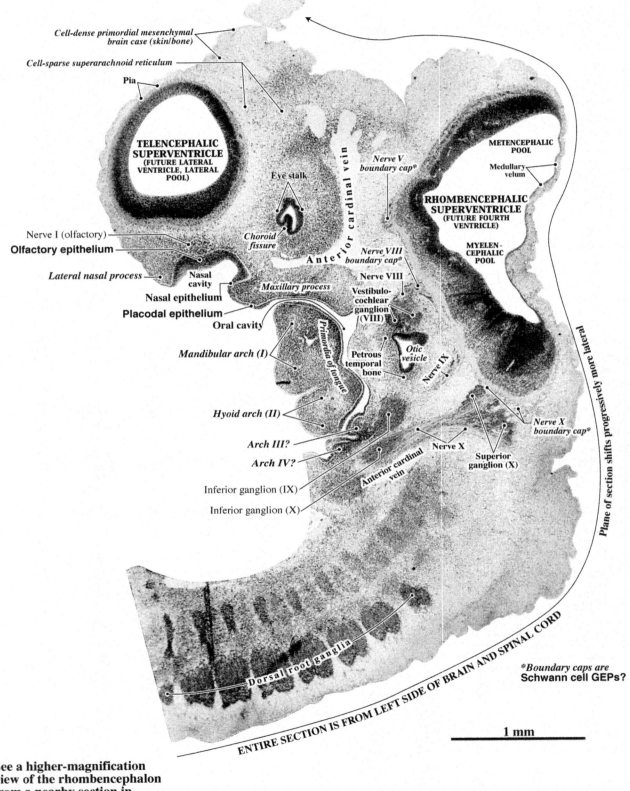

*Cell-dense primordial mesenchymal
brain case (skin/bone)*

Cell-sparse superarachnoid reticulum

Pia

**TELENCEPHALIC
SUPERVENTRICLE**
(FUTURE LATERAL
VENTRICLE, LATERAL
POOL)

Eye stalk

METENCEPHALIC
POOL

Medullary
velum

*Nerve V
boundary cap**

**RHOMBENCEPHALIC
SUPERVENTRICLE**
(FUTURE FOURTH
VENTRICLE)

*Choroid
fissure*

Nerve I (olfactory)

Olfactory epithelium

Anterior cardinal vein

A n t e r i o r c a r d i n a l v e i n

*Nerve VIII
boundary cap**

MYELENCEPHALIC
POOL

Nerve VIII

Lateral nasal process

Nasal
cavity

Nasal epithelium

Placodal epithelium

Maxillary process

Oral cavity

Vestibulocochlear
ganglion
(VIII)

Mandibular arch (I)

Primordia of tongue

Petrous
temporal
bone

*Otic
vesicle*

Nerve IX

Hyoid arch (II)

Arch III?

Arch IV?

Inferior ganglion (IX)

Inferior ganglion (X)

*Anterior cardinal
vein*

Nerve X

*Nerve X
boundary cap**

Superior
ganglion (X)

Plane of section shifts progressively more lateral

Dorsal root ganglia

ENTIRE SECTION IS FROM LEFT SIDE OF BRAIN AND SPINAL CORD

**Boundary caps are
Schwann cell GEPs?*

1 mm

**See a higher-magnification
view of the rhombencephalon
from a nearby section in
Plates 45A and B.**

Labeled on this page:
Central neural structures

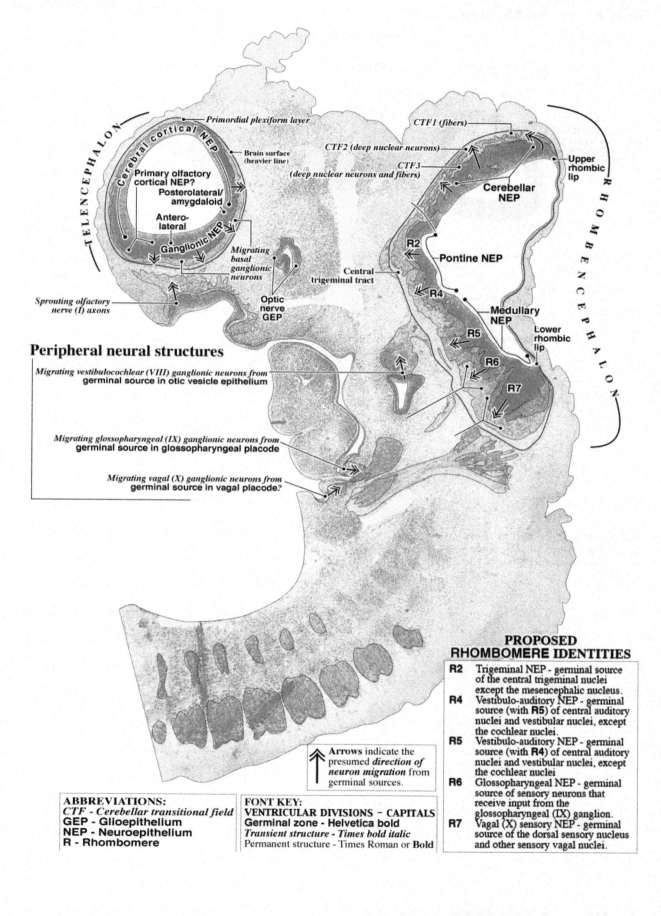

Primordial plexiform layer

Cerebral cortical NEP

Brain surface (heavier line)

Primary olfactory cortical NEP?

Posterolateral/ amygdaloid

Antero- lateral

Ganglionic NEP

Migrating basal ganglionic neurons

TELENCEPHALON

Sprouting olfactory nerve (I) axons

Optic nerve GEP

CTF1 *(fibers)*

CTF2 *(deep nuclear neurons)*

CTF3 *(deep nuclear neurons and fibers)*

Cerebellar NEP

Upper rhombic lip

R2

Pontine NEP

Central trigeminal tract

R4

Medullary NEP

Lower rhombic lip

R5

R6

R7

RHOMBENCEPHALON

Peripheral neural structures

Migrating vestibulocochlear (VIII) ganglionic neurons from germinal source in otic vesicle epithelium

Migrating glossopharyngeal (IX) ganglionic neurons from germinal source in glossopharyngeal placode

Migrating vagal (X) ganglionic neurons from germinal source in vagal placode?

Arrows indicate the presumed *direction of neuron migration* from germinal sources.

PROPOSED RHOMBOMERE IDENTITIES

R2 Trigeminal NEP - germinal source of the central trigeminal nuclei except the mesencephalic nucleus.

R4 Vestibulo-auditory NEP - germinal source (with **R5**) of central auditory nuclei and vestibular nuclei, except the cochlear nuclei.

R5 Vestibulo-auditory NEP - germinal source (with **R4**) of central auditory nuclei and vestibular nuclei, except the cochlear nuclei.

R6 Glossopharyngeal NEP - germinal source of sensory neurons that receive input from the glossopharyngeal (IX) ganglion.

R7 Vagal (X) sensory NEP - germinal source of the dorsal sensory nucleus and other sensory vagal nuclei.

ABBREVIATIONS:
CTF - Cerebellar transitional field
GEP - Glioepithelium
NEP - Neuroepithelium
R - Rhombomere

FONT KEY:
VENTRICULAR DIVISIONS – CAPITALS
Germinal zone - Helvetica bold
Transient structure - Times bold italic
Permanent structure - Times Roman or **Bold**

**PLATE 37A
CR 10.5 mm, GW 6.5, C6516
Slide 6, Section 15 Sagittal**

Labeled on this page:
Peripheral neural and non-neural
structures, brain ventricular divisions

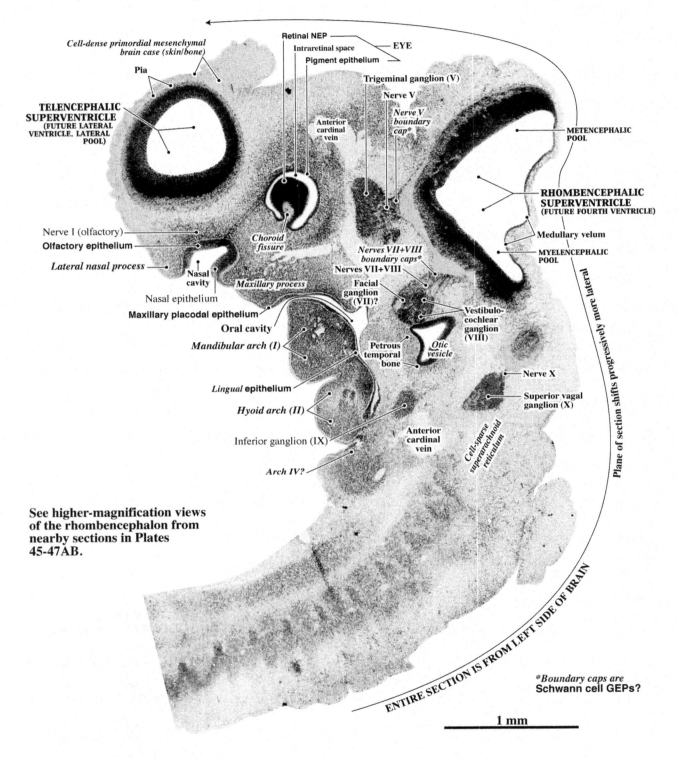

*Cell-dense primordial mesenchymal
brain case (skin/bone)*

Pia

Retinal NEP

Intraretinal space

Pigment epithelium

EYE

Trigeminal ganglion (V)

Nerve V

*Nerve V
boundary
cap**

**TELENCEPHALIC
SUPERVENTRICLE
(FUTURE LATERAL
VENTRICLE, LATERAL
POOL)**

Anterior
cardinal
vein

METENCEPHALIC
POOL

**RHOMBENCEPHALIC
SUPERVENTRICLE
(FUTURE FOURTH VENTRICLE)**

Nerve I (olfactory)

Olfactory epithelium

Lateral nasal process

*Choroid
fissure*

Nasal
cavity

Maxillary process

Nasal epithelium

Maxillary placodal epithelium

Oral cavity

Mandibular arch (I)

Lingual epithelium

Hyoid arch (II)

Inferior ganglion (IX)

Arch IV?

*Nerves VII+VIII
boundary caps**

Nerves VII+VIII

Facial
ganglion
(VII)?

Vestibulo-
cochlear
ganglion
(VIII)

Petrous
temporal
bone

*Otic
vesicle*

Anterior
cardinal
vein

Medullary velum

MYELENCEPHALIC
POOL

Nerve X

Superior vagal
ganglion (X)

*Cell-sparse
superarachnoid
reticulum*

Plane of section shifts progressively more lateral

**See higher-magnification views
of the rhombencephalon from
nearby sections in Plates
45-47AB.**

ENTIRE SECTION IS FROM LEFT SIDE OF BRAIN

**Boundary caps are
Schwann cell GEPs?*

1 mm

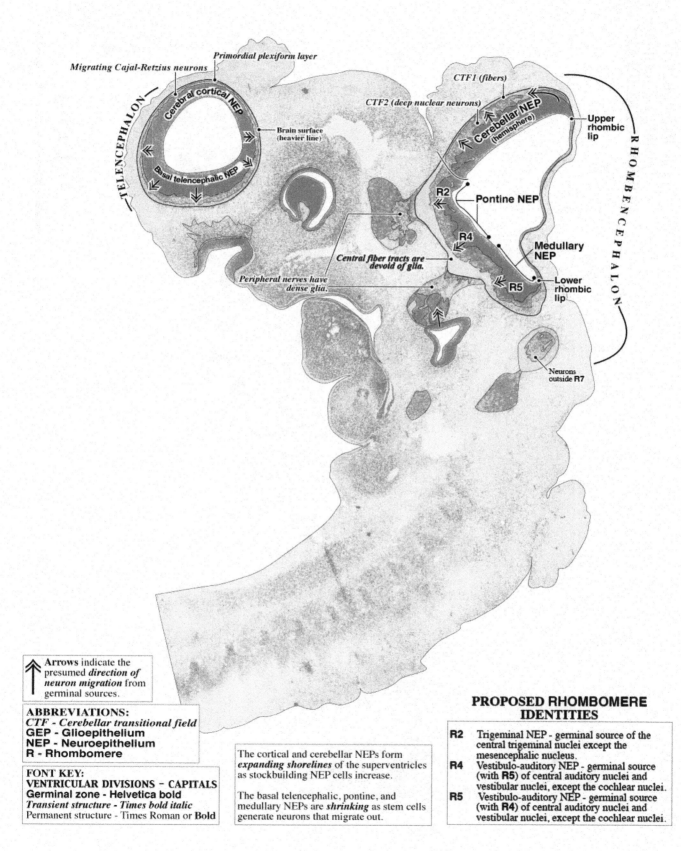

Migrating Cajal-Retzius neurons

Primordial plexiform layer

CTF1 (fibers)

CTF2 (deep nuclear neurons)

Cerebral cortical NEP

Cerebellar NEP (hemisphere)

Upper rhombic lip

Brain surface (heavier line)

TELENCEPHALON

Basal telencephalic NEP

R2

Pontine NEP

R4

RHOMBENCEPHALON

Medullary NEP

R5

Lower rhombic lip

Central fiber tracts are devoid of glia.

Peripheral nerves have dense glia.

Neurons outside R7

Arrows indicate the presumed *direction of neuron migration* from germinal sources.

ABBREVIATIONS:
CTF - Cerebellar transitional field
GEP - Glioepithelium
NEP - Neuroepithelium
R - Rhombomere

FONT KEY:
VENTRICULAR DIVISIONS – CAPITALS
Germinal zone - Helvetica bold
Transient structure - Times bold italic
Permanent structure - Times Roman or **Bold**

The cortical and cerebellar NEPs form *expanding shorelines* of the superventricles as stockbuilding NEP cells increase.

The basal telencephalic, pontine, and medullary NEPs are *shrinking* as stem cells generate neurons that migrate out.

PROPOSED RHOMBOMERE IDENTITIES

R2	Trigeminal NEP - germinal source of the central trigeminal nuclei except the mesencephalic nucleus.
R4	Vestibulo-auditory NEP - germinal source (with **R5**) of central auditory nuclei and vestibular nuclei, except the cochlear nuclei.
R5	Vestibulo-auditory NEP - germinal source (with **R4**) of central auditory nuclei and vestibular nuclei, except the cochlear nuclei.

98

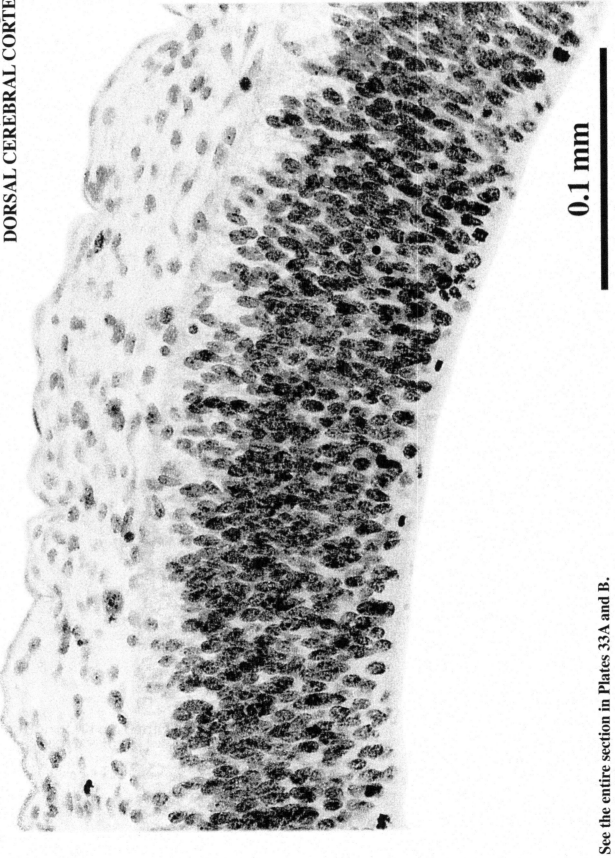

PLATE 38A

CR 10.5 mm, GW 6.5, C6516
Slide 11, Section 6 Sagittal
DORSAL CEREBRAL CORTEX

0.1 mm

See the entire section in Plates 33A and B.

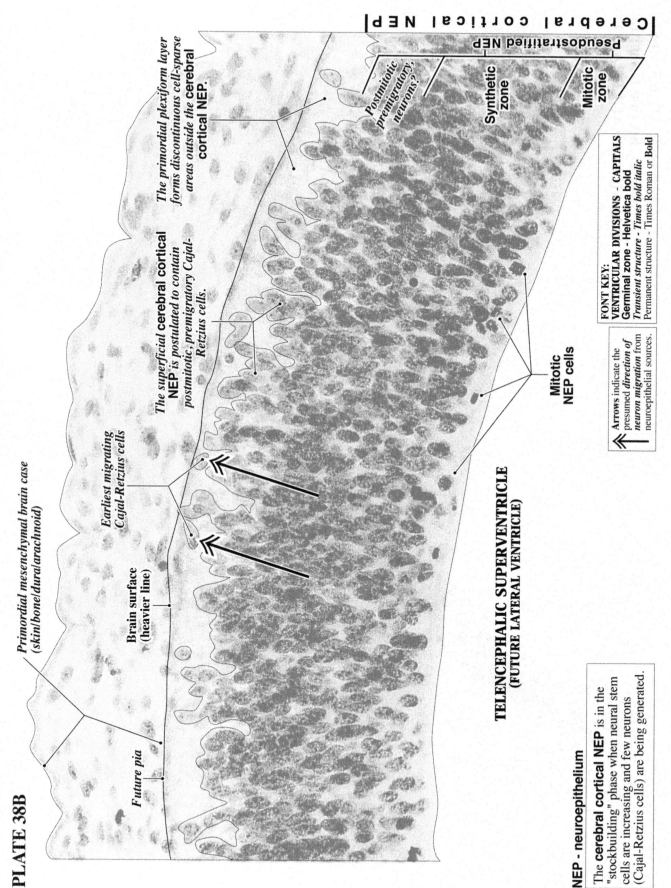

PLATE 38B

Primordial mesenchymal brain case (skin/bone/dura/arachnoid)

Future pia

Brain surface *(heavier line)*

Earliest migrating *Cajal-Retzius cells*

The superficial cerebral cortical **NEP** *is postulated to contain postmitotic, premigratory Cajal-Retzius cells.*

The primordial plexiform layer forms discontinuous cell-sparse areas outside the cerebral cortical NEP.

Postmitotic, premigratory, premigratory neurons?

Synthetic zone

Mitotic zone

Pseudostratified NEP

[Cerebral cortical NEP]

Mitotic NEP cells

TELENCEPHALIC SUPERVENTRICLE (FUTURE LATERAL VENTRICLE)

Arrows indicate the presumed *direction of neuron migration* from neuroepithelial sources.

FONT KEY:
VENTRICULAR DIVISIONS - CAPITALS
Germinal zone - Helvetica bold
Transient structure - Times bold italic
Permanent structure - Times Roman or **Bold**

NEP - neuroepithelium

The **cerebral cortical NEP** is in the "stockbuilding" phase when neural stem cells are increasing and few neurons (Cajal-Retzius cells) are being generated.

100

PLATE 39A

CR 10.5 mm, GW 6.5, C6516
Slide 11, Section 6 Sagittal
BASAL TELENCEPHALON

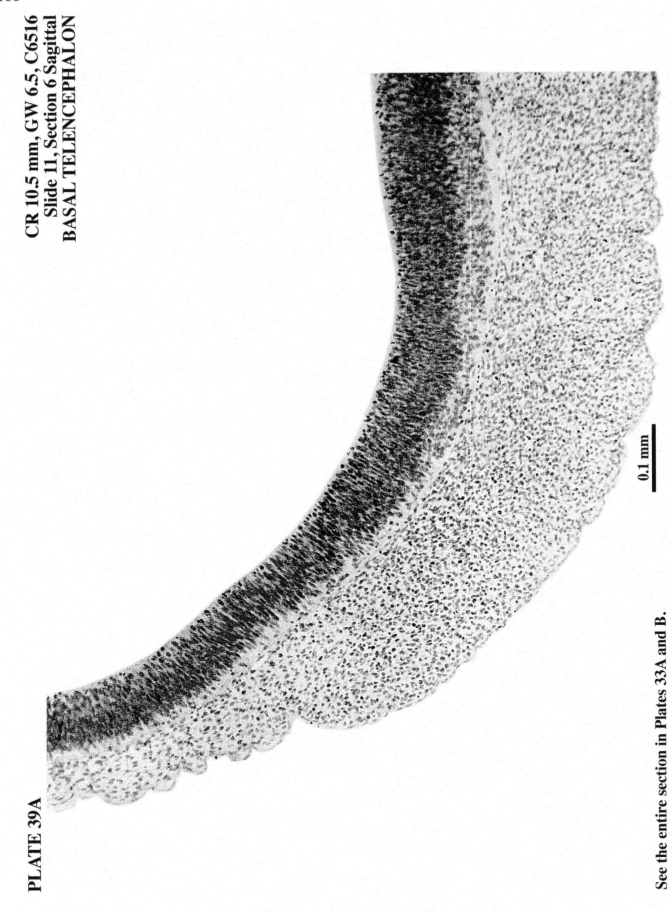

0.1 mm

See the entire section in Plates 33A and B.

101

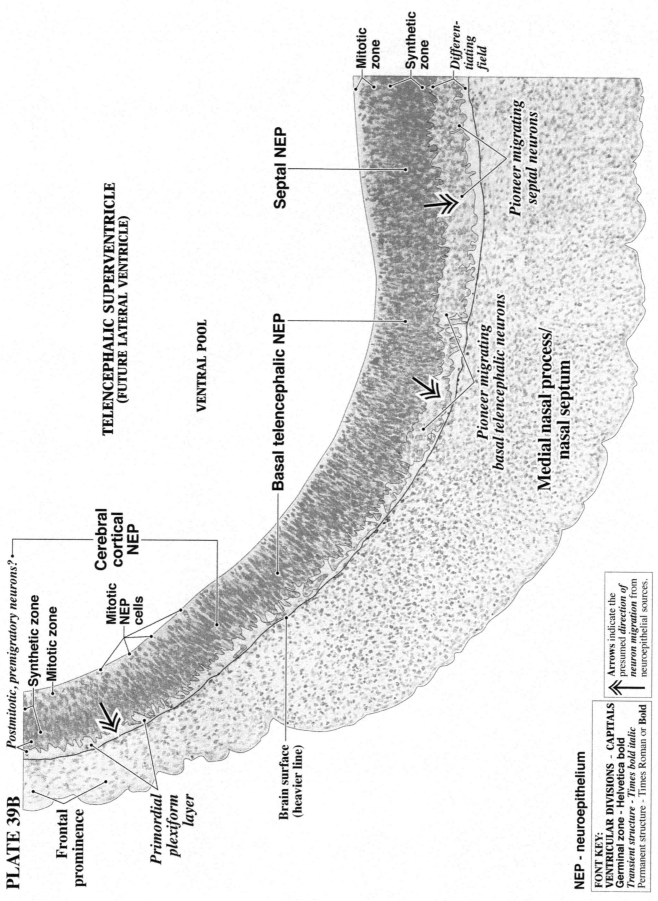

PLATE 39B

Postmitotic, premigratory neurons?

Synthetic zone
Mitotic zone

Mitotic NEP cells

Frontal prominence

Cerebral cortical NEP

Primordial plexiform layer

Brain surface (heavier line)

Basal telencephalic NEP

TELENCEPHALIC SUPERVENTRICLE
(FUTURE LATERAL VENTRICLE)

VENTRAL POOL

Septal NEP

Mitotic zone
Synthetic zone
Differentiating field

Pioneer migrating septal neurons

Pioneer migrating basal telencephalic neurons

Medial nasal process/ nasal septum

Arrows indicate the presumed *direction of neuron migration* from neuroepithelial sources.

NEP - neuroepithelium

FONT KEY:
VENTRICULAR DIVISIONS - CAPITALS
Germinal zone - Helvetica bold
Transient structure - Times bold italic
Permanent structure - Times Roman or **Bold**

PLATE 40A

CR 10.5 mm, GW 6.5, C6516
Slide 11, Section 6 Sagittal
SEPTUM/DIENCEPHALON

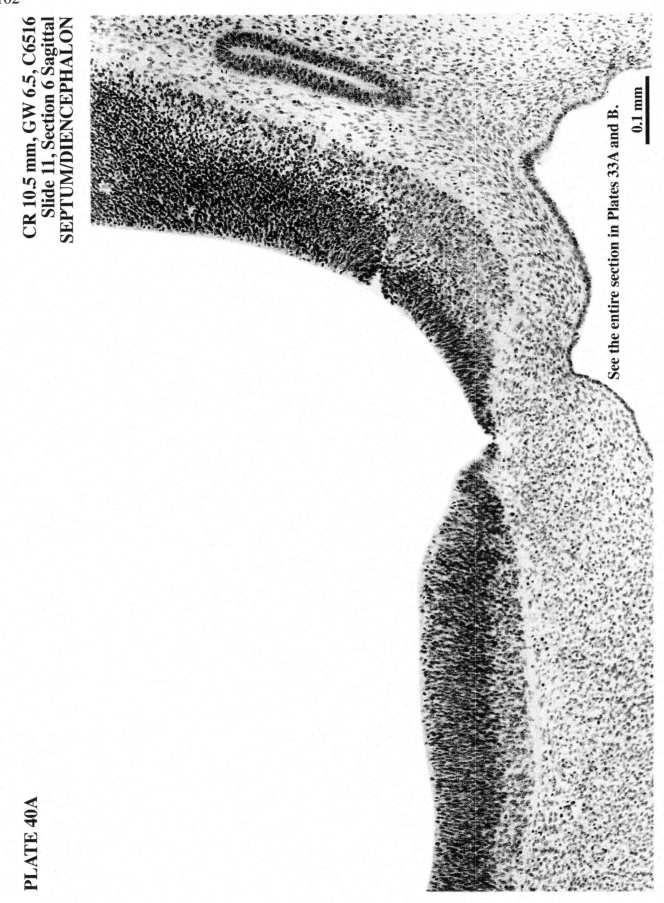

See the entire section in Plates 33A and B.

0.1 mm

103

PLATE 40B

NEP - neuroepithelium

FONT KEY:
VENTRICULAR DIVISIONS - CAPITALS
Germinal zone - Helvetica bold
Transient structure - Times bold italic
Permanent structure - Times Roman or Bold

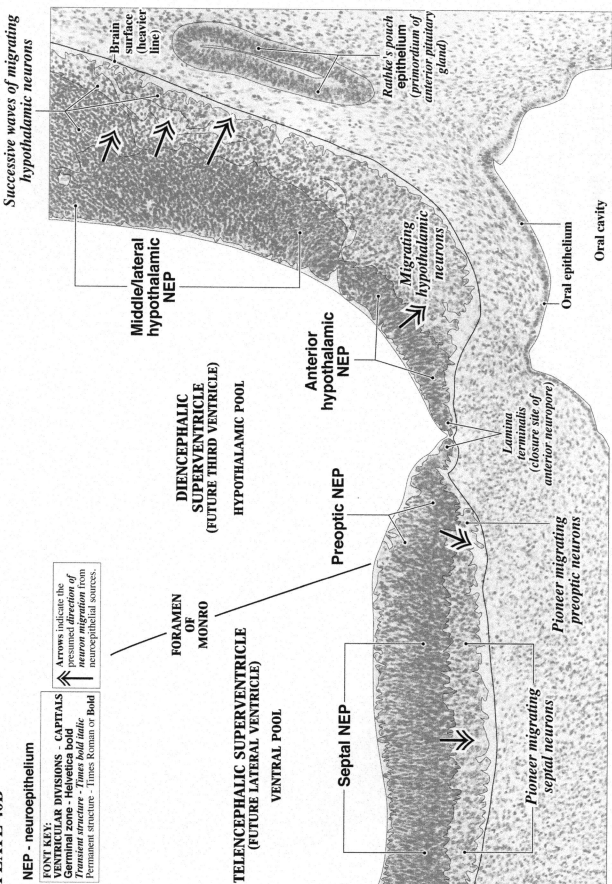

Successive waves of migrating hypothalamic neurons

Brain surface (heavier line)

Rathke's pouch epithelium (primordium of anterior pituitary gland)

Middle/lateral hypothalamic NEP

Oral cavity

Oral epithelium

Migrating hypothalamic neurons

DIENCEPHALIC SUPERVENTRICLE (FUTURE THIRD VENTRICLE)

HYPOTHALAMIC POOL

Anterior hypothalamic NEP

Lamina terminalis (closure site of anterior neuropore)

Preoptic NEP

Pioneer migrating preoptic neurons

Arrows indicate the presumed direction of neuron migration from neuroepithelial sources.

FORAMEN OF MONRO

TELENCEPHALIC SUPERVENTRICLE (FUTURE LATERAL VENTRICLE)

VENTRAL POOL

Septal NEP

Pioneer migrating septal neurons

104

PLATE 41A

CR 10.5 mm, GW 6.5, C6516
Slide 11, Section 6 Sagittal
MIDBRAIN TEGMENTUM

See the entire section in Plates 33A and B.

0.1 mm

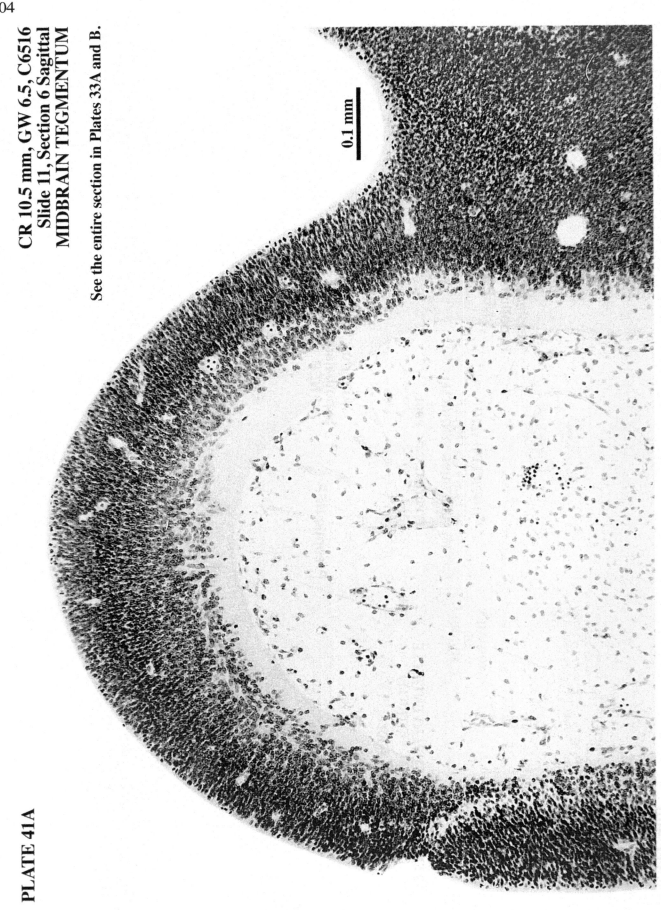

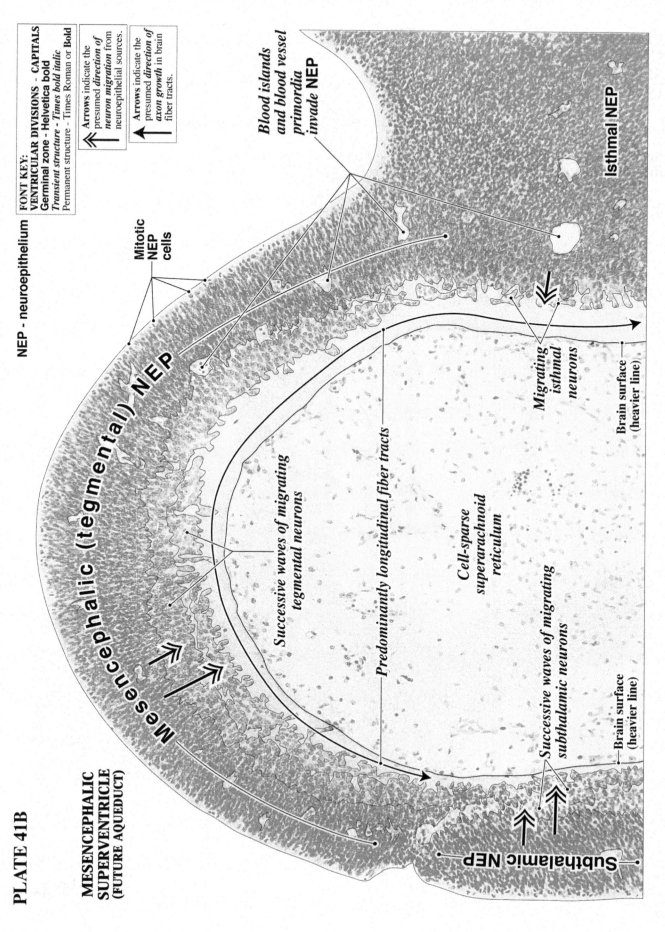

PLATE 41B

MESENCEPHALIC
SUPERVENTRICLE
(FUTURE AQUEDUCT)

NEP - neuroepithelium

FONT KEY:
VENTRICULAR DIVISIONS - CAPITALS
Germinal zone - Helvetica bold
Transient structure - Times bold italic
Permanent structure - Times Roman or **Bold**

⇐ Arrows indicate the presumed *direction of neuron migration* from neuroepithelial sources.

← Arrows indicate the presumed *direction of axon growth* in brain fiber tracts.

Blood islands
and blood vessel
primordia
invade NEP

Isthmal NEP

Mitotic
NEP
cells

Mesencephalic (tegmental) NEP

Migrating
isthmal
neurons

Brain surface
(heavier line)

Successive waves of migrating tegmental neurons

Predominantly longitudinal fiber tracts

Cell-sparse
superarachnoid
reticulum

Successive waves of migrating subthalamic neurons

Brain surface
(heavier line)

Subthalamic NEP

106

0.1 mm

PLATE 42B

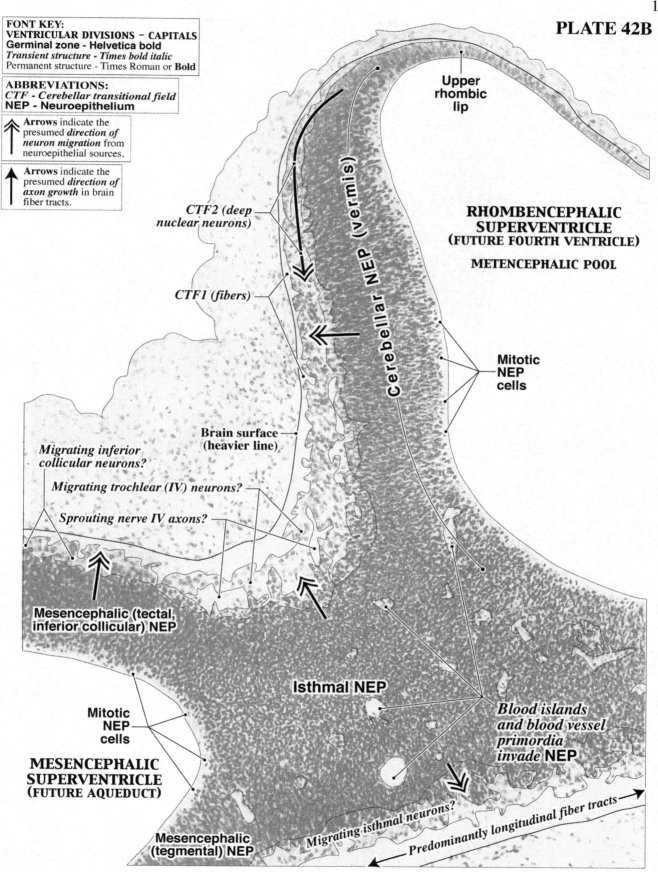

FONT KEY:
VENTRICULAR DIVISIONS – CAPITALS
Germinal zone - Helvetica bold
Transient structure - Times bold italic
Permanent structure - Times Roman or **Bold**

ABBREVIATIONS:
CTF - Cerebellar transitional field
NEP - Neuroepithelium

Arrows indicate the presumed *direction of neuron migration* from neuroepithelial sources.

Arrows indicate the presumed *direction of axon growth* in brain fiber tracts.

Upper rhombic lip

RHOMBENCEPHALIC
SUPERVENTRICLE
(FUTURE FOURTH VENTRICLE)

METENCEPHALIC POOL

CTF2 (deep nuclear neurons)

Cerebellar NEP (vermis)

CTF1 (fibers)

Mitotic NEP cells

Brain surface (heavier line)

Migrating inferior collicular neurons?

Migrating trochlear (IV) neurons?

Sprouting nerve IV axons?

Mesencephalic (tectal, inferior collicular) NEP

Isthmal NEP

Blood islands and blood vessel primordia invade NEP

Mitotic NEP cells

MESENCEPHALIC
SUPERVENTRICLE
(FUTURE AQUEDUCT)

Migrating isthmal neurons?

Predominantly longitudinal fiber tracts

Mesencephalic (tegmental) NEP

PLATE 43A

CR 10.5 mm, GW 6.5, C6516
Slide 8, Section 10 Sagittal
PONS/MEDULLA

PROPOSED RHOMBOMERE IDENTITIES

R2 Trigeminal NEP - germinal source of the central trigeminal nuclei except the mesencephalic nucleus.

R3 Facial sensory NEP - germinal source of sensory neurons that receive input from the facial (VII) ganglion.

R4 Vestibulo-auditory NEP - germinal source (with **R5**) of central auditory nuclei and vestibular nuclei, except the cochlear nuclei.

R5 Vestibulo-auditory NEP - germinal source (with **R4**) of central auditory nuclei and vestibular nuclei, except the cochlear nuclei.

R6 Glossopharyngeal NEP - germinal source of sensory neurons that receive input from the glossopharyngeal (IX) ganglion.

R7 Vagal (X) sensory NEP - germinal source of the dorsal sensory nucleus and other sensory vagal nuclei.

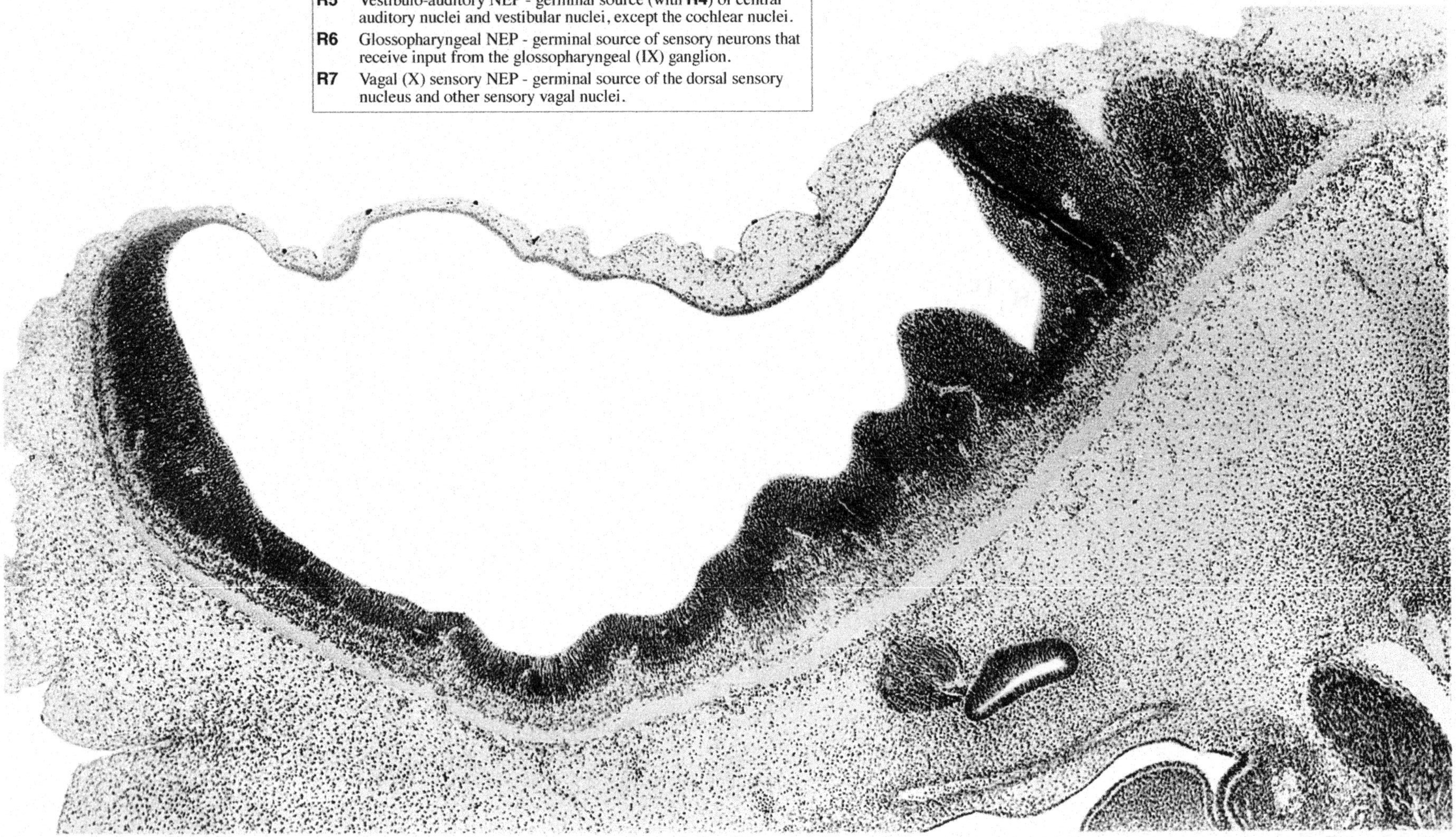

0.5 mm

A higher magnification view of the R2 to R7 neuroepithelium is in Plates 44A and B.

See an entire nearby section in Plates 35A and B.

109

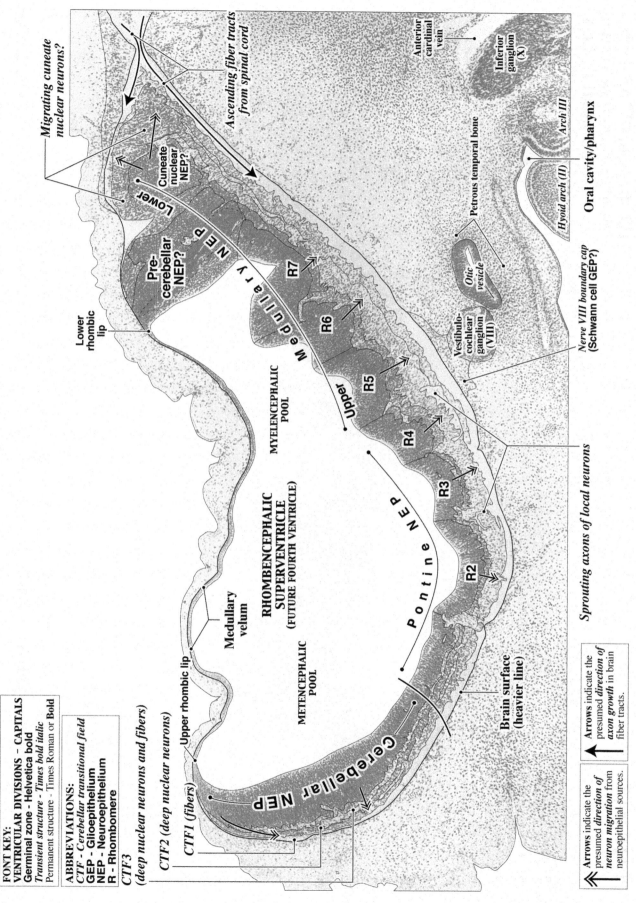

PLATE 43B

FONT KEY:
VENTRICULAR DIVISIONS – CAPITALS
Germinal zone – Helvetica bold
Transient structure – Times bold italic
Permanent structure – Times Roman or Bold

ABBREVIATIONS:
CTF – Cerebellar transitional field
GEP – Glioepithelium
NEP – Neuroepithelium
R – Rhombomere

CTF3 (deep nuclear neurons and fibers)

CTF2 (deep nuclear neurons)

CTF1 (fibers)

⇐ Arrows indicate the presumed *direction of neuron migration* from neuroepithelial sources.

← Arrows indicate the presumed *direction of axon growth* in brain fiber tracts.

Sprouting axons of local neurons

Migrating cuneate nuclear neurons?

Ascending fiber tracts from spinal cord

Anterior cardinal vein

Inferior ganglion (X)

Arch III

Oral cavity/pharynx

Hyoid arch (II)

Nerve VIII boundary cap (Schwann cell GEP?)

Petrous temporal bone

Otic vesicle

Vestibulocochlear ganglion (VIII)

Cuneate nuclear NEP?

Lower

Pre-cerebellar NEP?

Medullary NEP

Lower rhombic lip

R7
R6
R5
R4
R3
R2

Upper

Pontine NEP

MYELENCEPHALIC POOL

RHOMBENCEPHALIC SUPERVENTRICLE (FUTURE FOURTH VENTRICLE)

METENCEPHALIC POOL

Medullary velum

Upper rhombic lip

Cerebellar NEP

Brain surface (heavier line)

PLATE 44A

CR 10.5 mm, GW 6.5, C6516
Slide 8, Section 10 Sagittal
PONS/MEDULLA

PROPOSED RHOMBOMERE IDENTITIES

R2 Trigeminal NEP - germinal source of the central trigeminal nuclei except the mesencephalic nucleus.

R3 Facial sensory NEP - germinal source of sensory neurons that receive input from the facial (VII) ganglion.

R4 Vestibulo-auditory NEP - germinal source (with **R5**) of central auditory nuclei and vestibular nuclei, except the cochlear nuclei.

R5 Vestibulo-auditory NEP - germinal source (with **R4**) of central auditory nuclei and vestibular nuclei, except the cochlear nuclei.

R6 Glossopharyngeal NEP - germinal source of sensory neurons that receive input from the glossopharyngeal (IX) ganglion.

R7 Vagal (X) sensory NEP - germinal source of the dorsal sensory nucleus and other sensory vagal nuclei.

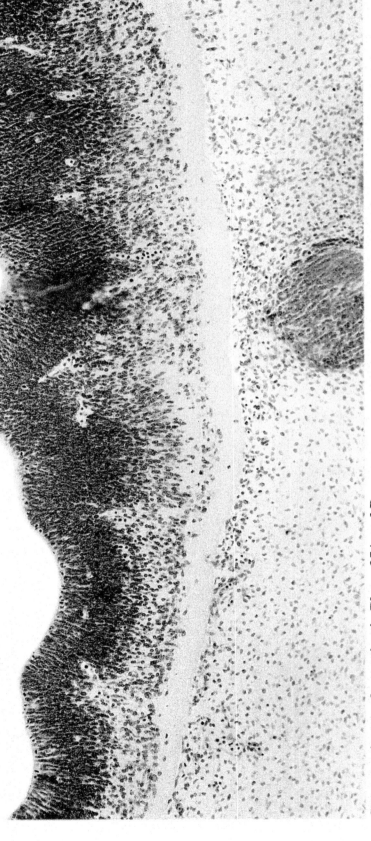

0.1 mm

See an entire nearby section in Plates 35A and B.

111

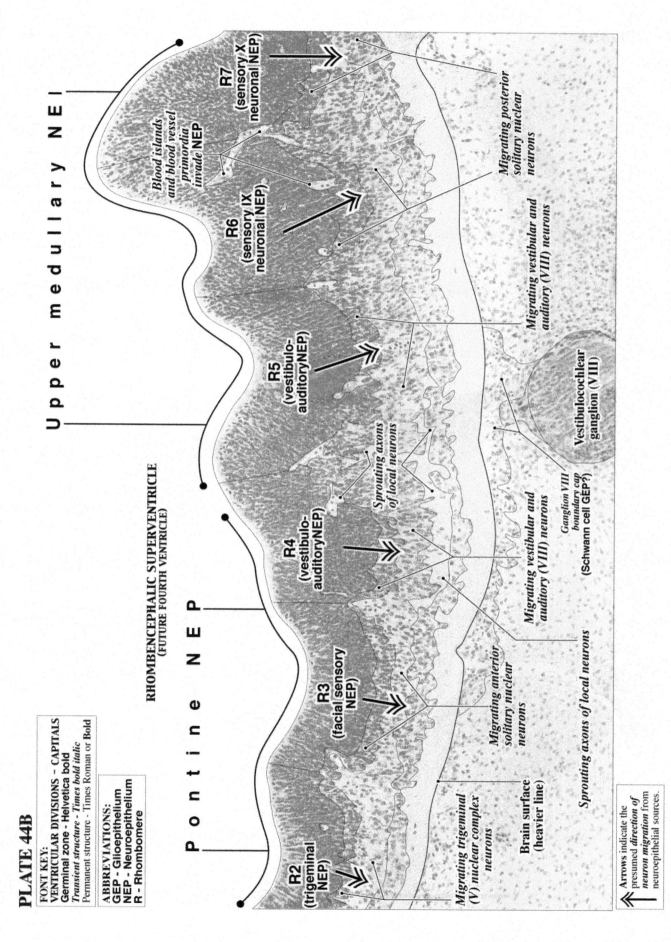

PLATE 44B

FONT KEY:
VENTRICULAR DIVISIONS – CAPITALS
Germinal zone – Helvetica bold
Transient structure – Times bold italic
Permanent structure – Times Roman or Bold

ABBREVIATIONS:
GEP - Glioepithelium
NEP - Neuroepithelium
R - Rhombomere

Upper medullary NEP

RHOMBENCEPHALIC SUPERVENTRICLE
(FUTURE FOURTH VENTRICLE)

Pontine NEP

R7
(sensory X
neuronal NEP)

R6
(sensory IX
neuronal NEP)

*Blood islands
and blood vessel
primordia* invade NEP

R5
(vestibulo-
auditoryNEP)

R4
(vestibulo-
auditoryNEP)

R3
(facial sensory NEP)

R2
(trigeminal NEP)

Migrating posterior solitary nuclear neurons

Migrating vestibular and auditory (VIII) neurons

Vestibulocochlear ganglion (VIII)

Sprouting axons of local neurons

Migrating vestibular and auditory (VIII) neurons

Ganglion VIII boundary cap (Schwann cell GEP?)

Migrating anterior solitary nuclear neurons

Sprouting axons of local neurons

Migrating trigeminal (V) nuclear complex neurons

Brain surface (heavier line)

Arrows indicate the presumed *direction of neuron migration* from neuroepithelial sources.

PLATE 45A

CR 10.5 mm, GW 6.5, C6516
Slide 7, Section 6 Sagittal
RHOMBENCEPHALON

Section is between Plates 36 and 37

PROPOSED **RHOMBOMERE** IDENTITIES

R2 Trigeminal NEP - germinal source of the central trigeminal nuclei except the mesencephalic nucleus.

R4 Vestibulo-auditory NEP - germinal source (with **R5**) of central auditory nuclei and vestibular nuclei, except the cochlear nuclei.

R5 Vestibulo-auditory NEP - germinal source (with **R4**) of central auditory nuclei and vestibular nuclei, except the cochlear nuclei.

R6 Glossopharyngeal NEP - germinal source of sensory neurons that receive input from the glossopharyngeal (IX) ganglion.

R7 Vagal (X) sensory NEP - germinal source of the dorsal sensory nucleus and other sensory vagal nuclei.

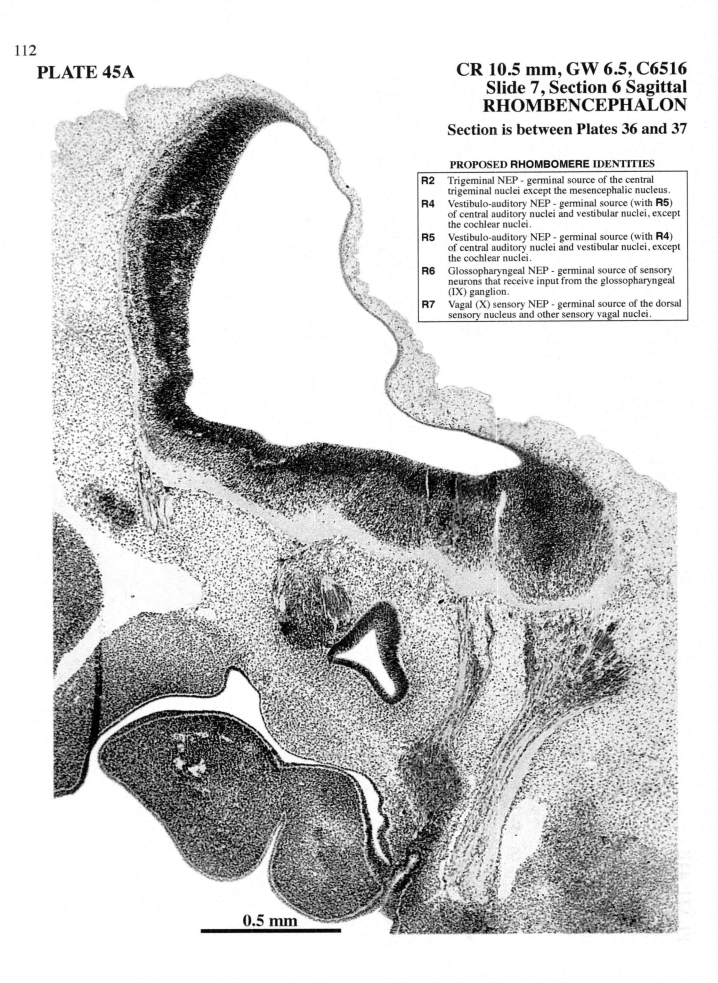

0.5 mm

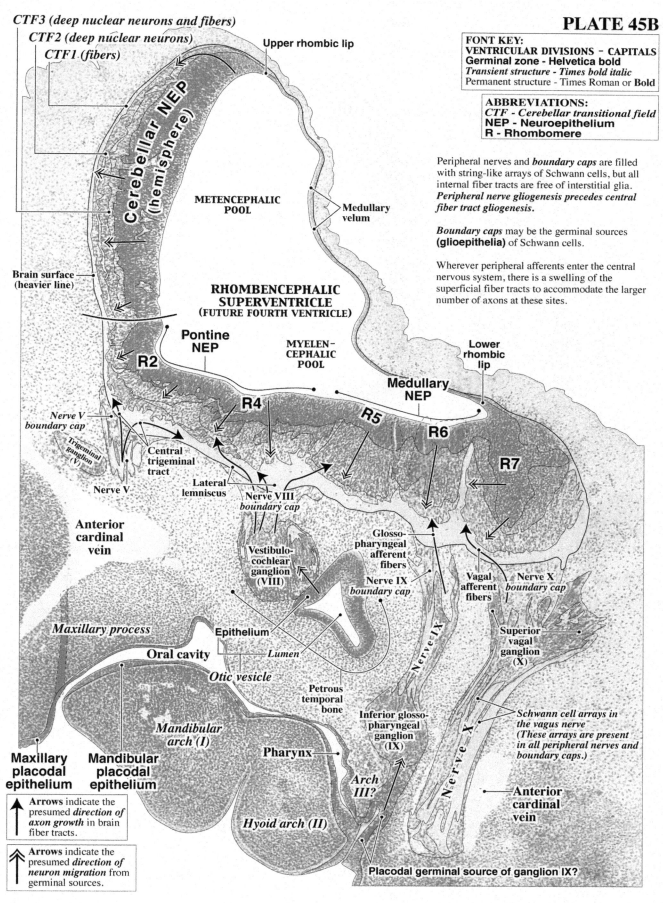

PLATE 45B

CTF3 *(deep nuclear neurons and fibers)*
CTF2 *(deep nuclear neurons)*
CTF1 *(fibers)*

Upper rhombic lip

FONT KEY:
VENTRICULAR DIVISIONS – CAPITALS
Germinal zone - Helvetica bold
Transient structure - Times bold italic
Permanent structure - Times Roman or **Bold**

ABBREVIATIONS:
CTF - Cerebellar transitional field
NEP - Neuroepithelium
R - Rhombomere

Cerebellar NEP
(hemisphere)

METENCEPHALIC POOL

Medullary velum

Peripheral nerves and *boundary caps* are filled with string-like arrays of Schwann cells, but all internal fiber tracts are of interstitial glia. *Peripheral nerve gliogenesis precedes central fiber tract gliogenesis.*

Boundary caps may be the germinal sources **(glioepithelia)** of Schwann cells.

Wherever peripheral afferents enter the central nervous system, there is a swelling of the superficial fiber tracts to accommodate the larger number of axons at these sites.

Brain surface (heavier line)

RHOMBENCEPHALIC SUPERVENTRICLE
(FUTURE FOURTH VENTRICLE)

Pontine NEP

MYELEN-CEPHALIC POOL

Lower rhombic lip

R2

R4

R5

Medullary NEP

R6

R7

Nerve V boundary cap

Trigeminal ganglion (V)

Central trigeminal tract

Lateral lemniscus

Nerve VIII boundary cap

Nerve V

Glosso-pharyngeal afferent fibers

Vagal afferent fibers

Nerve X boundary cap

Anterior cardinal vein

Vestibulo-cochlear ganglion (VIII)

Nerve IX boundary cap

Superior vagal ganglion (X)

Maxillary process

Epithelium

Lumen

Oral cavity

Otic vesicle

Mandibular arch (I)

Petrous temporal bone

Inferior glosso-pharyngeal ganglion (IX)

Nerve IX

Nerve X

Schwann cell arrays in the vagus nerve (These arrays are present in all peripheral nerves and boundary caps.)

Maxillary placodal epithelium

Mandibular placodal epithelium

Pharynx

Arch III?

Anterior cardinal vein

Hyoid arch (II)

Arrows indicate the presumed *direction of axon growth* in brain fiber tracts.

Arrows indicate the presumed *direction of neuron migration* from germinal sources.

Placodal germinal source of ganglion IX?

PLATE 46A

CR 10.5 mm, GW 6.5, C6516
Slide 6, Section 11 Sagittal
RHOMBENCEPHALON
See a nearby section
in Plates 37A and B.

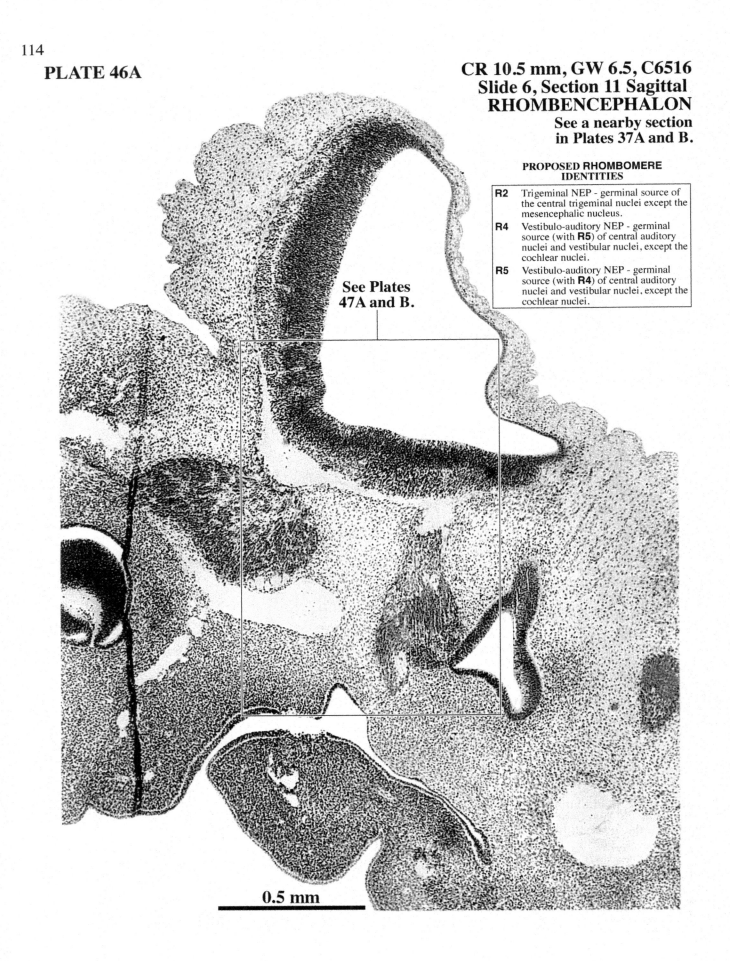

**PROPOSED RHOMBOMERE
IDENTITIES**

R2	Trigeminal NEP - germinal source of the central trigeminal nuclei except the mesencephalic nucleus.
R4	Vestibulo-auditory NEP - germinal source (with **R5**) of central auditory nuclei and vestibular nuclei, except the cochlear nuclei.
R5	Vestibulo-auditory NEP - germinal source (with **R4**) of central auditory nuclei and vestibular nuclei, except the cochlear nuclei.

See Plates
47A and B.

0.5 mm

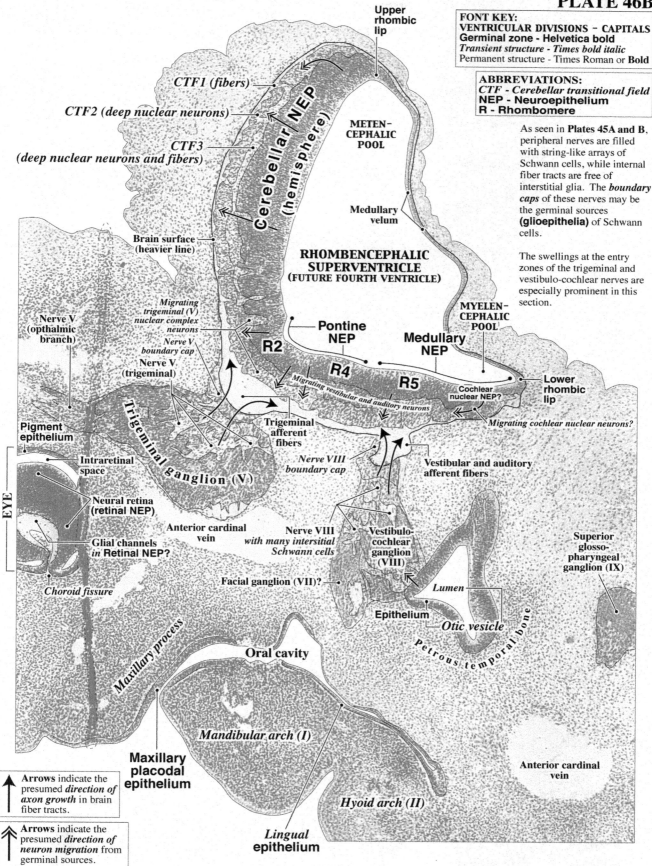

FONT KEY:
VENTRICULAR DIVISIONS – CAPITALS
Germinal zone - Helvetica bold
Transient structure - Times bold italic
Permanent structure - Times Roman or **Bold**

ABBREVIATIONS:
CTF - Cerebellar transitional field
NEP - Neuroepithelium
R - Rhombomere

As seen in **Plates 45A and B**, peripheral nerves are filled with string-like arrays of Schwann cells, while internal fiber tracts are free of interstitial glia. The *boundary caps* of these nerves may be the germinal sources **(glioepithelia)** of Schwann cells.

The swellings at the entry zones of the trigeminal and vestibulo-cochlear nerves are especially prominent in this section.

Upper rhombic lip

CTF1 (fibers)

CTF2 (deep nuclear neurons)

CTF3
(deep nuclear neurons and fibers)

Cerebellar NEP *(hemisphere)*

METEN-CEPHALIC POOL

Medullary velum

RHOMBENCEPHALIC SUPERVENTRICLE (FUTURE FOURTH VENTRICLE)

Brain surface (heavier line)

Migrating trigeminal (V) nuclear complex neurons

Nerve V boundary cap

Nerve V (trigeminal)

Nerve V (opthalmic branch)

Pontine NEP

R2

R4

R5

MYELEN-CEPHALIC POOL

Medullary NEP

Cochlear nuclear NEP?

Lower rhombic lip

Migrating vestibular and auditory neurons

Migrating cochlear nuclear neurons?

Pigment epithelium

Trigeminal ganglion (V)

Trigeminal afferent fibers

Nerve VIII boundary cap

Vestibular and auditory afferent fibers

Intraretinal space

Neural retina (retinal NEP)

Anterior cardinal vein

Nerve VIII *with many interstitial Schwann cells*

Vestibulo-cochlear ganglion (VIII)

Superior glosso-pharyngeal ganglion (IX)

EYE

Glial channels *in* **Retinal NEP?**

Facial ganglion (VII)?

Lumen

Epithelium

Otic vesicle

Petrous temporal bone

Choroid fissure

Maxillary process

Oral cavity

Anterior cardinal vein

Maxillary placodal epithelium

Mandibular arch (I)

Hyoid arch (II)

Lingual epithelium

Arrows indicate the presumed *direction of axon growth* in brain fiber tracts.

Arrows indicate the presumed *direction of neuron migration* from germinal sources.

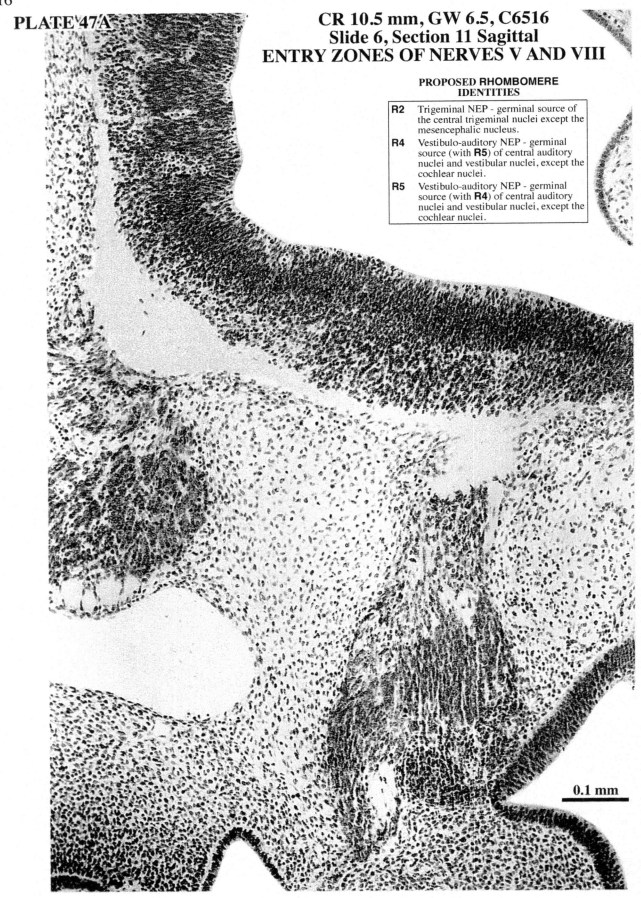

PLATE 47A

CR 10.5 mm, GW 6.5, C6516
Slide 6, Section 11 Sagittal
ENTRY ZONES OF NERVES V AND VIII

**PROPOSED RHOMBOMERE
IDENTITIES**

R2 Trigeminal NEP - germinal source of
the central trigeminal nuclei except the
mesencephalic nucleus.

R4 Vestibulo-auditory NEP - germinal
source (with **R5**) of central auditory
nuclei and vestibular nuclei, except the
cochlear nuclei.

R5 Vestibulo-auditory NEP - germinal
source (with **R4**) of central auditory
nuclei and vestibular nuclei, except the
cochlear nuclei.

0.1 mm

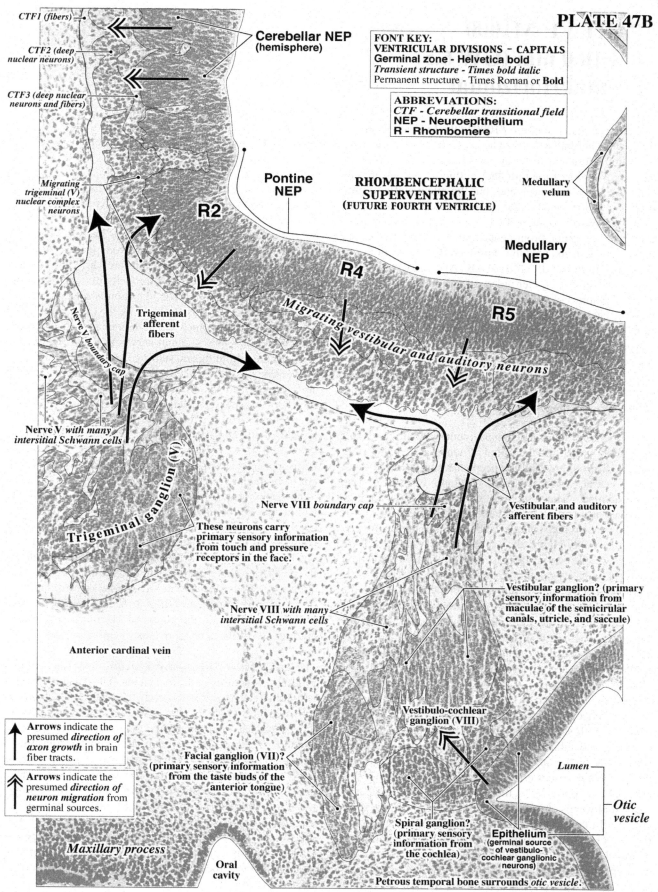

117

PLATE 47B

CTF1 (fibers)

CTF2 (deep nuclear neurons)

Cerebellar NEP (hemisphere)

CTF3 (deep nuclear neurons and fibers)

FONT KEY:
VENTRICULAR DIVISIONS – CAPITALS
Germinal zone - Helvetica bold
Transient structure - Times bold italic
Permanent structure - Times Roman or **Bold**

ABBREVIATIONS:
CTF - Cerebellar transitional field
NEP - Neuroepithelium
R - Rhombomere

Medullary velum

Migrating trigeminal (V) nuclear complex neurons

Pontine NEP

RHOMBENCEPHALIC SUPERVENTRICLE (FUTURE FOURTH VENTRICLE)

R2

Medullary NEP

R4

R5

Nerve V boundary cap

Trigeminal afferent fibers

Migrating vestibular and auditory neurons

Nerve V *with many intersitial Schwann cells*

Trigeminal ganglion (V)

Nerve VIII boundary cap

These neurons carry primary sensory information from touch and pressure receptors in the face.

Vestibular and auditory afferent fibers

Vestibular ganglion? (primary sensory information from maculae of the semicirular canals, utricle, and saccule)

Nerve VIII *with many intersitial Schwann cells*

Anterior cardinal vein

Vestibulo-cochlear ganglion (VIII)

Arrows indicate the presumed *direction of axon growth* in brain fiber tracts.

Arrows indicate the presumed *direction of neuron migration* from germinal sources.

Facial ganglion (VII)? (primary sensory information from the taste buds of the anterior tongue)

Lumen

Otic vesicle

Spiral ganglion? (primary sensory information from the cochlea)

Epithelium (germinal source of vestibulo-cochlear ganglionic neurons)

Maxillary process

Oral cavity

Petrous temporal bone surrounds *otic vesicle*.

PART VI: M1000
CR 10.0 mm (GW 6.5)
Frontal/Horizontal

This specimen is embryo #1000 in the Minot Collection, designated here as M1000. The crown-rump length (CR) is 10 mm estimated to be at gestational week (GW) 6.5. Most of M1000's forebrain and midbrain sections are cut (10-μm) in the coronal plane, but the plane shifts to predominantly horizontal in the posterior midbrain, pons, and medulla. We photographed 64 sections at low magnification from the frontal prominence to the posterior tips of the mesencephalon and medulla. Fourteen of these sections are illustrated in **Plates 48AB to 61AB**. All photographs were used to produce computer-aided 3-D reconstructions of the external features of M1000's brain and eye (**Figure 5**), and to show each illustrated section *in situ* (*insets*, **Plates 48A to 61A**). Each illustrated section shows the brain with all surrounding tissues. Labels in **A Plates** (normal-contrast images) identify non-neural and peripheral neural structures; labels in **B Plates** (low-contrast images) identify central neural structures. **Plate 62AB** shows high-magnification views of the telencephalic NEP.

All parts of the telencephalic neuroepithelium (NEP) are rapidly increasing their pool of neuronal and glial stem cells as they enlarge the telencephalic superventricle into paired lateral expansions. The cortical NEP is becoming more defined. Many sections show a denser apical part than the basal part. It is postulated that premigratory, postmitotic neurons are sequestered basally before moving out. To support this claim, the cortical primordial plexiform layer is nearly devoid of cells except in far ventrolateral areas. The basal ganglionic and basal telencephalic neuroepithelia have only a thin layer of adjacent migrating neurons. The posterior olfactory epithelium has only partially invaginated into a developing nasal cavity, while the anterior epithelium is a placode in the anterolateral head. Still, cellular densities outside the placode and invaginated epithelium may be supporting cells surrounding the first olfactory nerve fibers.

The diencephalic neuroepithelium surrounds a superventricle with shrinking shorelines in the preoptic, hypothalamic, and subthalamic areas. It is also postulated that basal parts of the anterior hypothalamic and subthalamic NEPs contain sequestered premigratory, postmitotic neurons. More posteriorly, these NEPs are surrounded by sequential waves of migrating neurons invading the parenchyma. In contrast, the thalamic neuroepithelium is still the "stockbuilding" stage, increasing its population of neuronal and glial stem cells as the thalamic pool of the diencephalic superventricle expands. The eye is now connected to the ventral diencephalon by a thick, short stalk and may contain a glioepithelium for the future optic nerve. The retinal NEP is clearly differentiated from the pigment epithelium.

The mesencephalon contains a stockbuilding NEP in the pretectum and tectum (virtually no adjacent migrating neurons outside the NEP). The tegmental and isthmal NEPs are thick, but their population of stem cells begins to decrease as massive waves of migrating neurons leave. Both the pons and medulla have NEPs that are shrinking as they unload their neuronal and glial progeny into an expanding parenchyma. In lateral areas, the rhombomere NEPS are less prominent. Many sections show the close association between the rhombomeres and boundary caps marking entry zones of sensory cranial nerves V, VII, VIII, IX, and X. Cells are migrating and settling in longitudinal arrays at the pontine flexure. Most regions of the pons and medulla are characterized by longitudinal bands of neurons interspersed with medial and lateral fiber bands. The cells have a few clumps that may be early nuclear divisions. We tentatively identify a superior olivary complex and trigeminal motor and sensory nuclei. Many are settling in the reticular formation throughout the pons and medulla, which does not have definite nuclear divisions even in mature brains. Some facial motor neurons are migrating from medial to lateral, leaving behind their axons in a small, but definite genu of the facial motor nerve. Migrating inferior olive neurons are in the posterior intramural migratory stream outside the precerebellar NEP in the posterior lower rhombic lip. The solitary nucleus and tract cannot be identified, although solitary nuclear neurons are undoubtedly migrating outside rhombomere NEPs 3, 6, and 7. The subpial fiber band is thick in the pons and medulla, especially at entry points of the sensory nerves.

The cerebellar NEP is busy producing deep nuclear neurons that are settling in distinct bands. The data in **Table 1B** would indicate that it is too early to see deep nuclear neurons outside the NEP. Our cerebellar neurogenetic timetables are based on the vermis; the hemispheres were not studied. Obviously, the layers in the cerebellar transitional field in M1000 and C8314 indicate that many deep neurons have been produced before the CR 10- to 10.5-mm stage and have migrated to their sojourn zones. We postulate that these are neurons that will settle in the hemispheres (dentate nucleus). The thick cerebellar NEP has a denser apical part than the basal part, because the base is postulated to contain sequestered postmitotic nurons. We postulate that many of these are Purkinje cells and younger deep nuclear neurons.

M1000 Computer-aided 3-D Brain Reconstructions

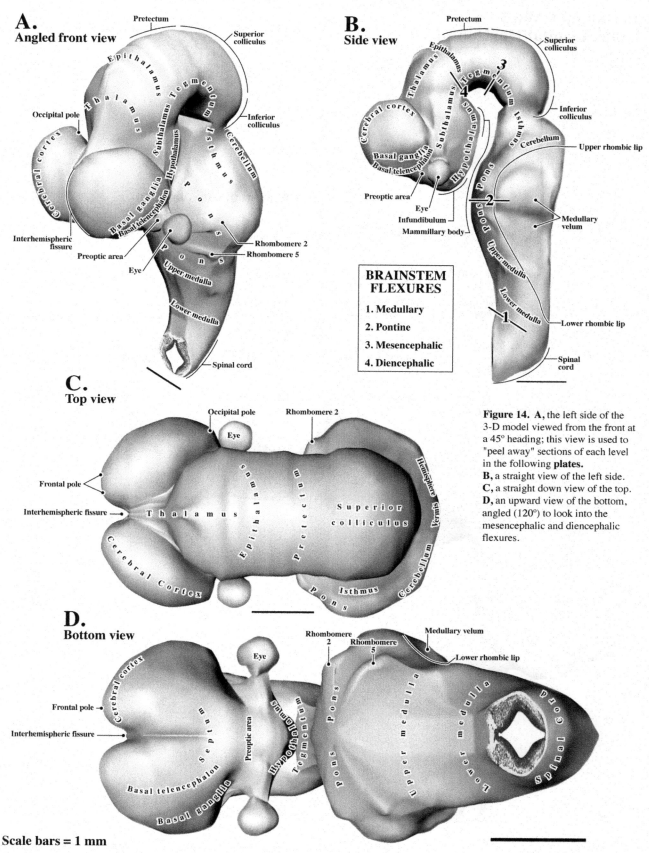

A. Angled front view

B. Side view

C. Top view

D. Bottom view

BRAINSTEM FLEXURES
1. Medullary
2. Pontine
3. Mesencephalic
4. Diencephalic

Figure 14. A, the left side of the 3-D model viewed from the front at a 45° heading; this view is used to "peel away" sections of each level in the following **plates.**
B, a straight view of the left side.
C, a straight down view of the top.
D, an upward view of the bottom, angled (120°) to look into the mesencephalic and diencephalic flexures.

Scale bars = 1 mm

PLATE 48A

CR 10 mm, GW6.5
M1000, Frontal/Horizontal
Section 29

Non-neural structures labeled

Interhemispheric fissure

Cell-sparse superarachnoid reticulum
(brain parenchymal expansion zone)

Cell-dense primordial
mesenchymal brain case
(skin/bone)

Future dura
(internal border
of brain case)

Pia

Olfactory placode

Section 29 brain *in situ*

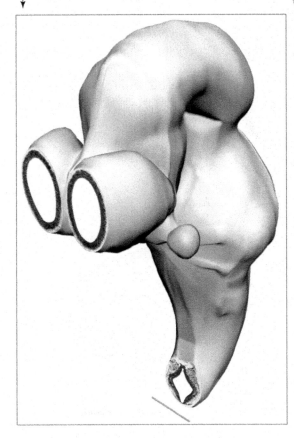

1 mm

Neural structures labeled

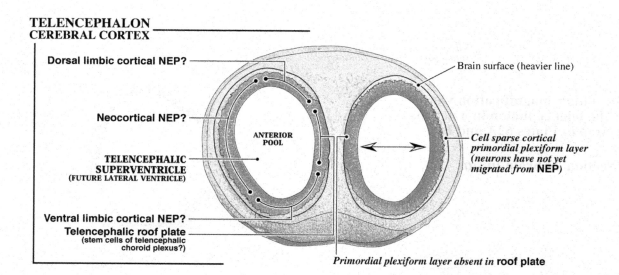

TELENCEPHALON
CEREBRAL CORTEX

Dorsal limbic cortical NEP? ——

Neocortical NEP? ——

TELENCEPHALIC
SUPERVENTRICLE
(FUTURE LATERAL VENTRICLE)

Ventral limbic cortical NEP? ——
Telencephalic roof plate ——
(stem cells of telencephalic
choroid plexus?)

ANTERIOR
POOL

Brain surface (heavier line)

*Cell sparse cortical
primordial plexiform layer
(neurons have not yet
migrated from NEP)*

Primordial plexiform layer absent in **roof plate**

NEP - Neuroepithelium

Arrows indicate the regionally
expanding shoreline of the
superventricle with increase in
stockbuilding NEP cells.

FONT KEY:
VENTRICULAR DIVISIONS - CAPITALS
Germinal zone - Helvetica bold
Transient structure - Times bold italic
Permanent structure - Times Roman or **Bold**

PLATE 49A

CR 10 mm, GW6.5
M1000, Frontal/Horizontal
Section 42

Peripheral neural and non-neural structures labeled

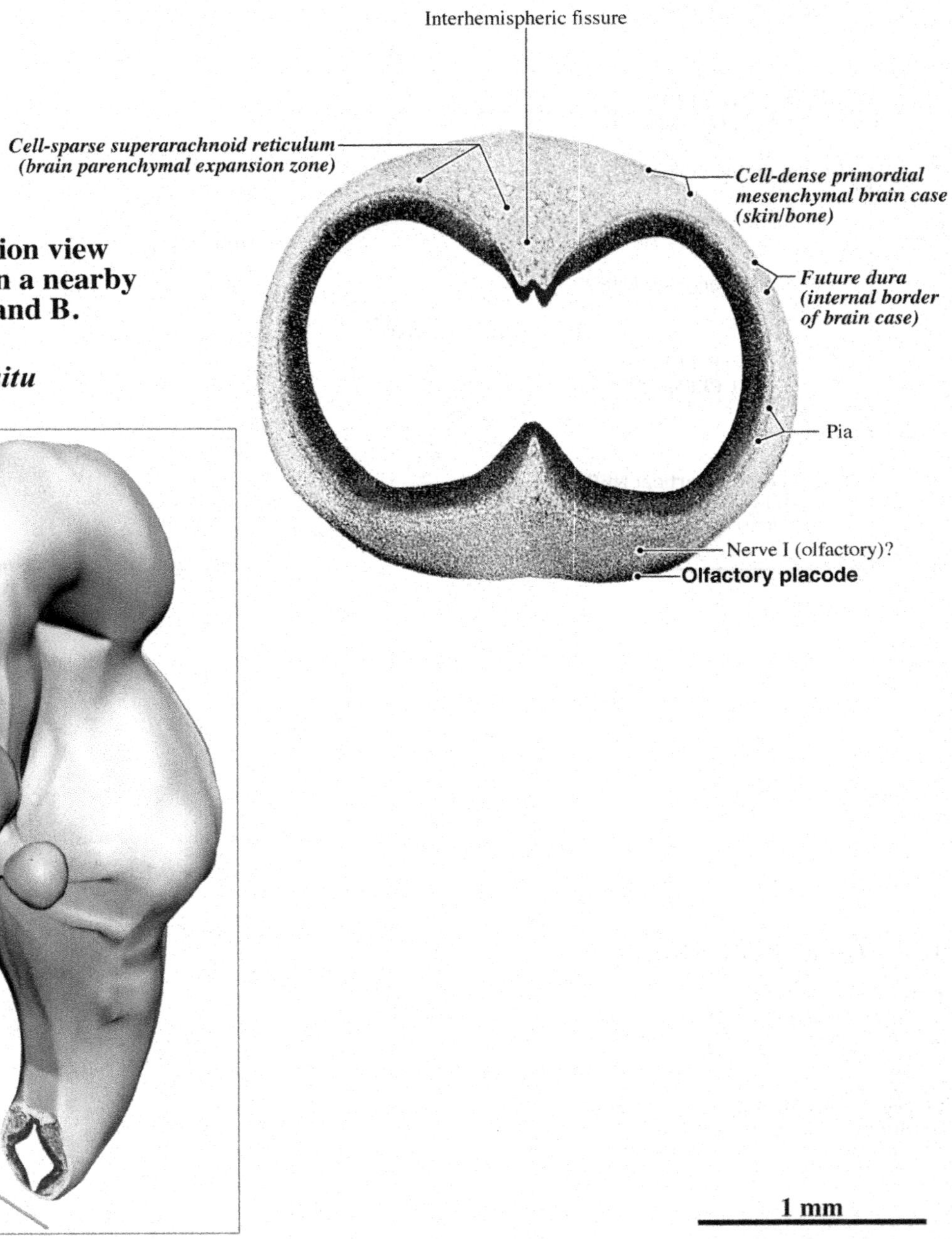

**See a high-magnification view
of the telencephalon in a nearby
section in Plates 62A and B.**

Interhemispheric fissure

Cell-sparse superarachnoid reticulum
(brain parenchymal expansion zone)

Cell-dense primordial
mesenchymal brain case
(skin/bone)

Future dura
(internal border
of brain case)

Pia

Nerve I (olfactory)?
Olfactory placode

Section 42 brain *in situ*

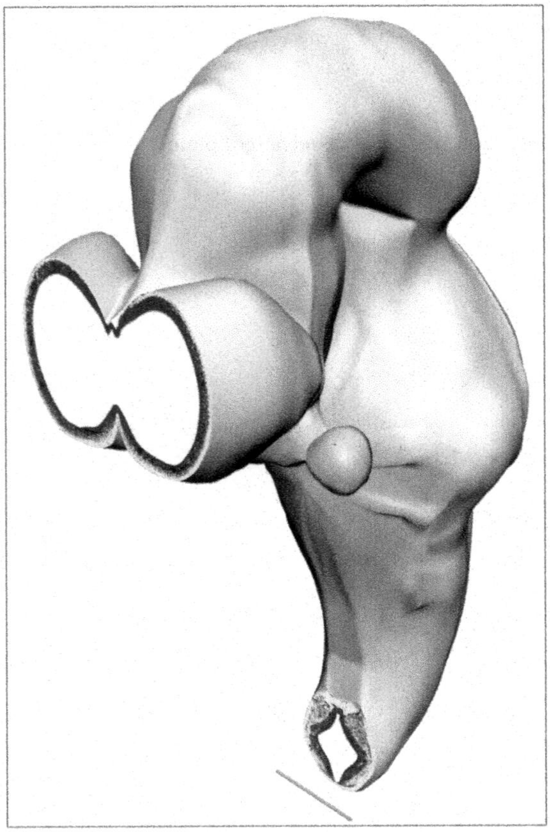

1 mm

Central neural structures labeled

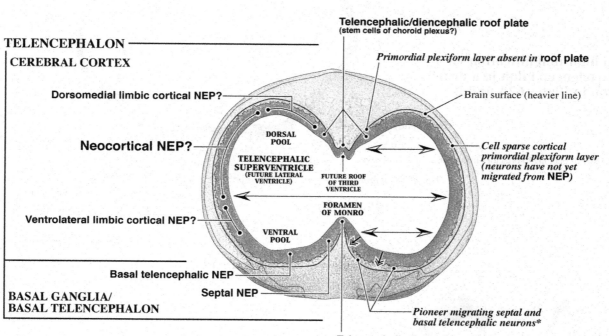

Telencephalic/diencephalic roof plate
(stem cells of choroid plexus?)

Primordial plexiform layer absent in **roof plate**

TELENCEPHALON

CEREBRAL CORTEX

Dorsomedial limbic cortical NEP?

Brain surface (heavier line)

Neocortical NEP?

DORSAL POOL

TELENCEPHALIC SUPERVENTRICLE
(FUTURE LATERAL VENTRICLE)

FUTURE ROOF OF THIRD VENTRICLE

Cell sparse cortical primordial plexiform layer (neurons have not yet migrated from **NEP***)*

FORAMEN OF MONRO

Ventrolateral limbic cortical NEP?

VENTRAL POOL

Basal telencephalic NEP

Septal NEP

Pioneer migrating septal and basal telencephalic neurons*

BASAL GANGLIA/ BASAL TELENCEPHALON

Telencephalic floor plate

*Note that the group of basal telencephalic migrating neurons may contain mitral cells heading for the future olfactory evagination.

NEP - Neuroepithelium

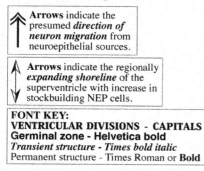

Arrows indicate the presumed *direction of neuron migration* from neuroepithelial sources.

Arrows indicate the regionally *expanding shoreline* of the superventricle with increase in stockbuilding NEP cells.

FONT KEY:
VENTRICULAR DIVISIONS - CAPITALS
Germinal zone - Helvetica bold
Transient structure - Times bold italic
Permanent structure - Times Roman or **Bold**

124

CR 10 mm, GW6.5
M1000, Frontal/Horizontal
Section 100

Peripheral neural and
non-neural structures labeled

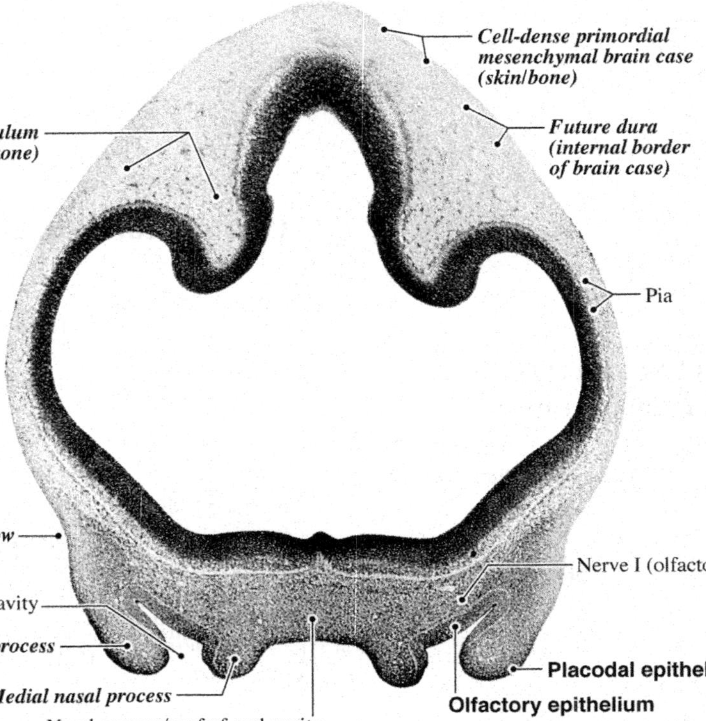

Cell-dense primordial
mesenchymal brain case
(skin/bone)

Future dura
(internal border
of brain case)

Cell-sparse superarachnoid reticulum
(brain parenchymal expansion zone)

Pia

See a high-magnification view
of the telencephalon in a nearby
section in Plates 62A and B.

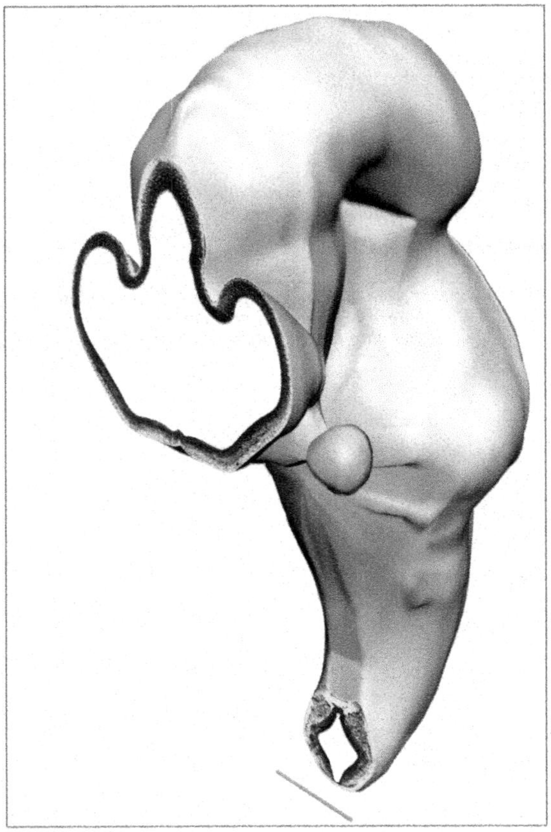

Naso-optic furrow

Nostril opening to olfactory invagination and nasal cavity

Lateral nasal process

Nerve I (olfactory)

Medial nasal process

Nasal septum/roof of oral cavity

Placodal epithelium
Olfactory epithelium

1 mm

Section 100 brain *in situ*

The GW5.5 Face and Neck
Figure 247C modified (Patten, 1953, p. 429.)

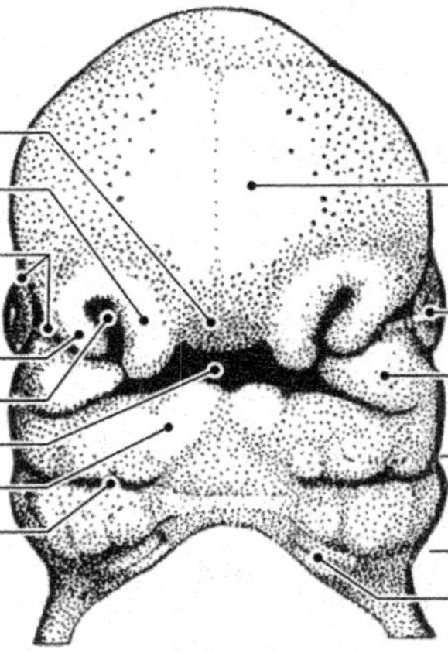

Nasal septum

Medial nasal process

Naso-optic furrow

Lateral nasal process

Nostril

Mouth

Mandible

Hyo-mandibular cleft

Frontal prominence

Eye

Maxillary process

Mandibular arch (I)

Hyoid arch (II)

Laryngeal
cartilages?

Central neural structures labeled

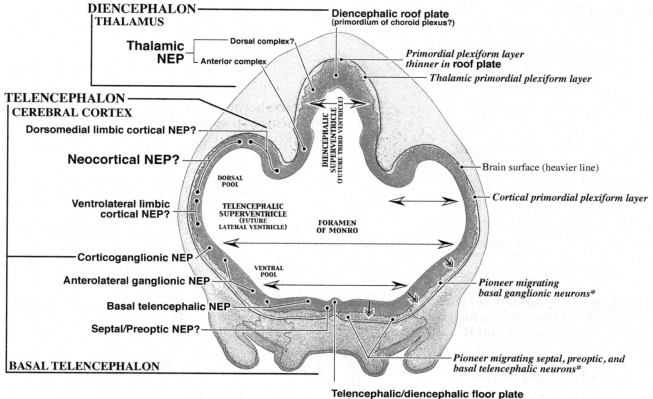

DIENCEPHALON
THALAMUS

Thalamic NEP
— Dorsal complex?
— Anterior complex

Diencephalic roof plate
(primordium of choroid plexus?)

*Primordial plexiform layer
thinner in* roof plate

Thalamic primordial plexiform layer

TELENCEPHALON
CEREBRAL CORTEX

Dorsomedial limbic cortical NEP?

Neocortical NEP?

Ventrolateral limbic cortical NEP?

Corticoganglionic NEP

Anterolateral ganglionic NEP

Basal telencephalic NEP

Septal/Preoptic NEP?

BASAL TELENCEPHALON

DORSAL POOL

DIENCEPHALIC SUPERVENTRICLE
(FUTURE THIRD VENTRICLE)

TELENCEPHALIC SUPERVENTRICLE
(FUTURE LATERAL VENTRICLE)

VENTRAL POOL

FORAMEN OF MONRO

Brain surface (heavier line)

Cortical primordial plexiform layer

*Pioneer migrating basal ganglionic neurons**

*Pioneer migrating septal, preoptic, and basal telencephalic neurons**

Telencephalic/diencephalic floor plate
(lamina terminalis in junction of preoptic area and septum?)

Note that some of these migrating neurons are mitral cells heading for the future olfactory bulb.

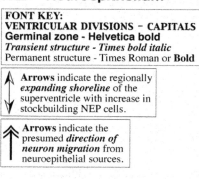

NEP - Neuroepithelium

FONT KEY:
VENTRICULAR DIVISIONS – CAPITALS
Germinal zone - Helvetica bold
Transient structure - Times bold italic
Permanent structure - Times Roman or **Bold**

Arrows indicate the regionally *expanding shoreline* of the superventricle with increase in stockbuilding NEP cells.

Arrows indicate the presumed *direction of neuron migration* from neuroepithelial sources.

126

PLATE 51A

CR 10 mm, GW6.5
M1000, Frontal/Horizontal
Section 128

**Peripheral neural and
non-neural structures labeled**

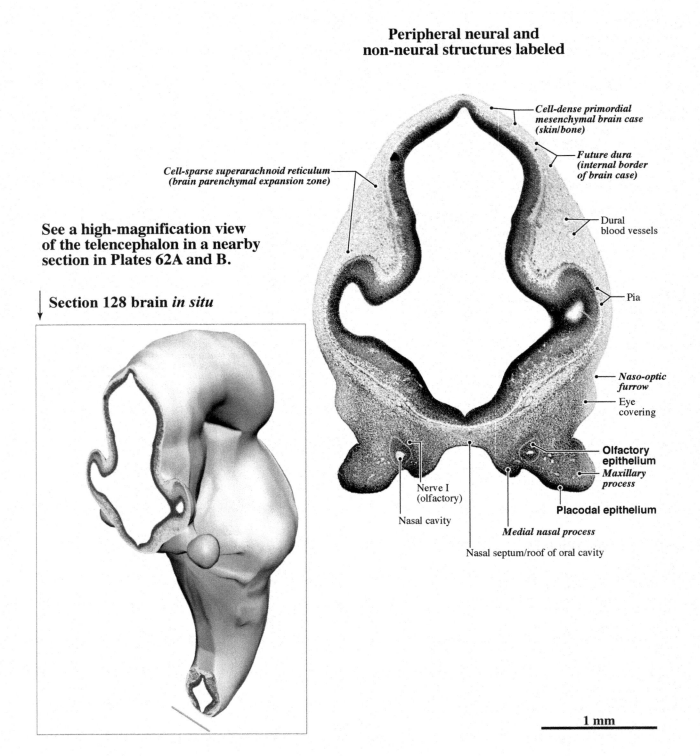

Cell-sparse superarachnoid reticulum
(brain parenchymal expansion zone)

*Cell-dense primordial
mesenchymal brain case
(skin/bone)*

*Future dura
(internal border
of brain case)*

Dural
blood vessels

Pia

*Naso-optic
furrow*

Eye
covering

**Olfactory
epithelium**
*Maxillary
process*

Placodal epithelium

Nerve I
(olfactory)

Nasal cavity

Medial nasal process

Nasal septum/roof of oral cavity

**See a high-magnification view
of the telencephalon in a nearby
section in Plates 62A and B.**

Section 128 brain *in situ*

1 mm

Central neural structures labeled

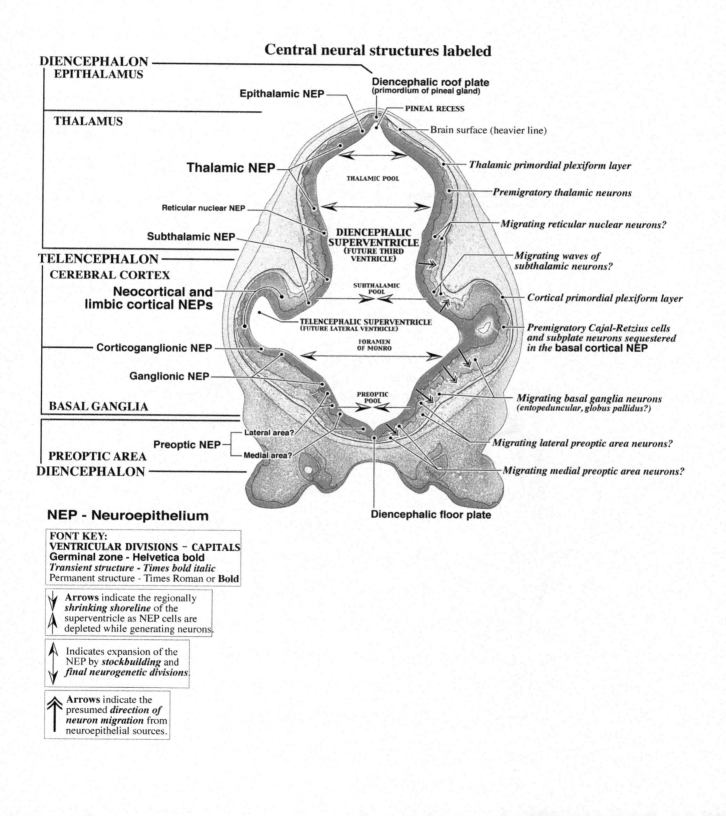

DIENCEPHALON
EPITHALAMUS

Epithalamic NEP

Diencephalic roof plate
(primordium of pineal gland)

PINEAL RECESS

THALAMUS

Brain surface (heavier line)

Thalamic NEP

Thalamic primordial plexiform layer

THALAMIC POOL

Premigratory thalamic neurons

Reticular nuclear NEP

Migrating reticular nuclear neurons?

Subthalamic NEP

**DIENCEPHALIC
SUPERVENTRICLE**
(FUTURE THIRD
VENTRICLE)

*Migrating waves of
subthalamic neurons?*

TELENCEPHALON
CEREBRAL CORTEX

SUBTHALAMIC
POOL

Cortical primordial plexiform layer

**Neocortical and
limbic cortical NEPs**

TELENCEPHALIC SUPERVENTRICLE
(FUTURE LATERAL VENTRICLE)

*Premigratory Cajal-Retzius cells
and subplate neurons sequestered
in the* **basal cortical NEP**

Corticoganglionic NEP

FORAMEN
OF MONRO

Ganglionic NEP

*Migrating basal ganglia neurons
(entopeduncular, globus pallidus?)*

BASAL GANGLIA

PREOPTIC
POOL

Lateral area?

Preoptic NEP

Medial area?

Migrating lateral preoptic area neurons?

PREOPTIC AREA
DIENCEPHALON

Migrating medial preoptic area neurons?

Diencephalic floor plate

NEP - Neuroepithelium

FONT KEY:
VENTRICULAR DIVISIONS – CAPITALS
Germinal zone - Helvetica bold
Transient structure - Times bold italic
Permanent structure - Times Roman or **Bold**

Arrows indicate the regionally
shrinking shoreline of the
superventricle as NEP cells are
depleted while generating neurons.

Indicates expansion of the
NEP by *stockbuilding* and
final neurogenetic divisions.

Arrows indicate the
presumed *direction of
neuron migration* from
neuroepithelial sources.

128

PLATE 52A

CR 10 mm, GW6.5
M1000, Frontal/Horizontal
Section 169

**Peripheral neural and
non-neural structures labeled**

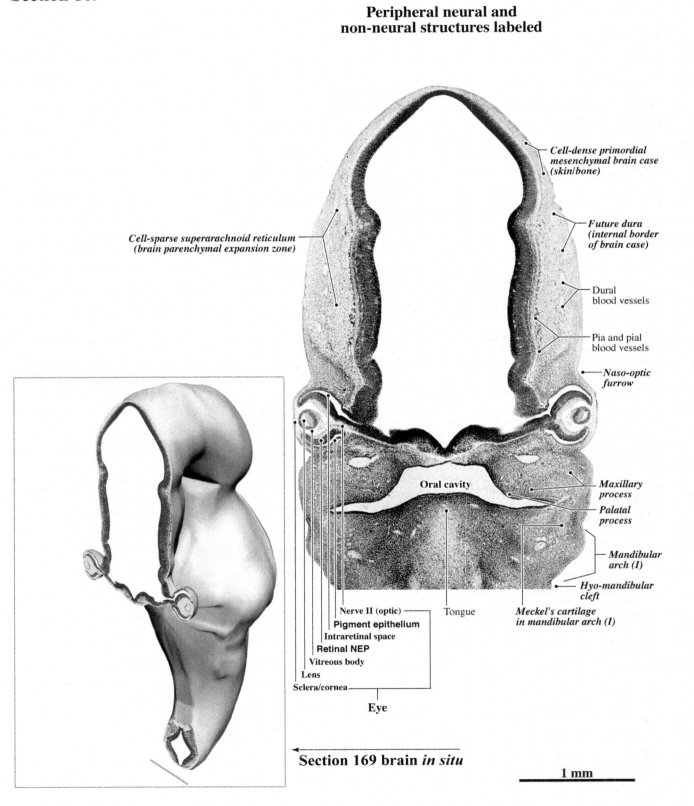

*Cell-dense primordial
mesenchymal brain case
(skin/bone)*

*Future dura
(internal border
of brain case)*

Dural
blood vessels

Pia and pial
blood vessels

*Cell-sparse superarachnoid reticulum
(brain parenchymal expansion zone)*

*Naso-optic
furrow*

Oral cavity

*Maxillary
process*

*Palatal
process*

*Mandibular
arch (I)*

*Hyo-mandibular
cleft*

Nerve II (optic)

Tongue

*Meckel's cartilage
in mandibular arch (I)*

Pigment epithelium

Intraretinal space

Retinal NEP

Vitreous body

Lens

Sclera/cornea

Eye

Section 169 brain *in situ*

1 mm

Central neural structures labeled

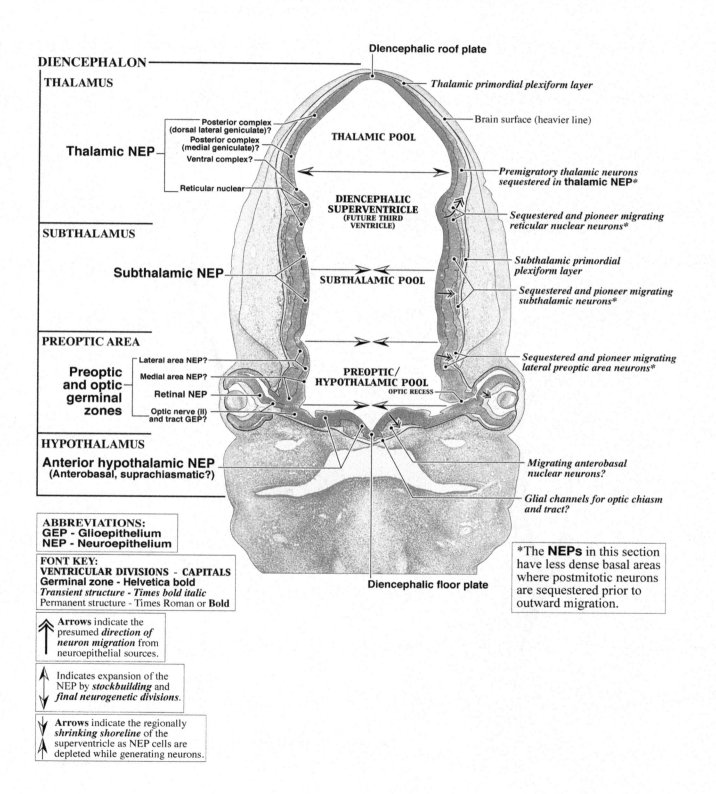

Diencephalic roof plate

DIENCEPHALON

THALAMUS

Thalamic primordial plexiform layer

Posterior complex
(dorsal lateral geniculate)?
Posterior complex
(medial geniculate)?
Ventral complex?

Thalamic NEP

Reticular nuclear

THALAMIC POOL

Brain surface (heavier line)

*Premigratory thalamic neurons
sequestered in* **thalamic NEP***

**DIENCEPHALIC
SUPERVENTRICLE**
(FUTURE THIRD
VENTRICLE)

*Sequestered and pioneer migrating
reticular nuclear neurons**

SUBTHALAMUS

Subthalamic NEP

SUBTHALAMIC POOL

*Subthalamic primordial
plexiform layer*

*Sequestered and pioneer migrating
subthalamic neurons**

PREOPTIC AREA

**Preoptic
and optic
germinal
zones**

Lateral area NEP?
Medial area NEP?
Retinal NEP
Optic nerve (II)
and tract GEP?

**PREOPTIC/
HYPOTHALAMIC POOL**
OPTIC RECESS

*Sequestered and pioneer migrating
lateral preoptic area neurons**

HYPOTHALAMUS

Anterior hypothalamic NEP
(Anterobasal, suprachiasmatic?)

*Migrating anterobasal
nuclear neurons?*

*Glial channels for optic chiasm
and tract?*

**ABBREVIATIONS:
GEP - Glioepithelium
NEP - Neuroepithelium**

**FONT KEY:
VENTRICULAR DIVISIONS - CAPITALS
Germinal zone - Helvetica bold
Transient structure - Times bold italic
Permanent structure - Times Roman or Bold**

Diencephalic floor plate

*The **NEPs** in this section
have less dense basal areas
where postmitotic neurons
are sequestered prior to
outward migration.

Arrows indicate the
presumed *direction of
neuron migration* from
neuroepithelial sources.

Indicates expansion of the
NEP by *stockbuilding* and
final neurogenetic divisions.

Arrows indicate the regionally
shrinking shoreline of the
superventricle as NEP cells are
depleted while generating neurons.

130

PLATE 53A

CR 10 mm, GW6.5
M1000, Frontal/Horizontal
Section 192

Peripheral neural and
non-neural structures labeled

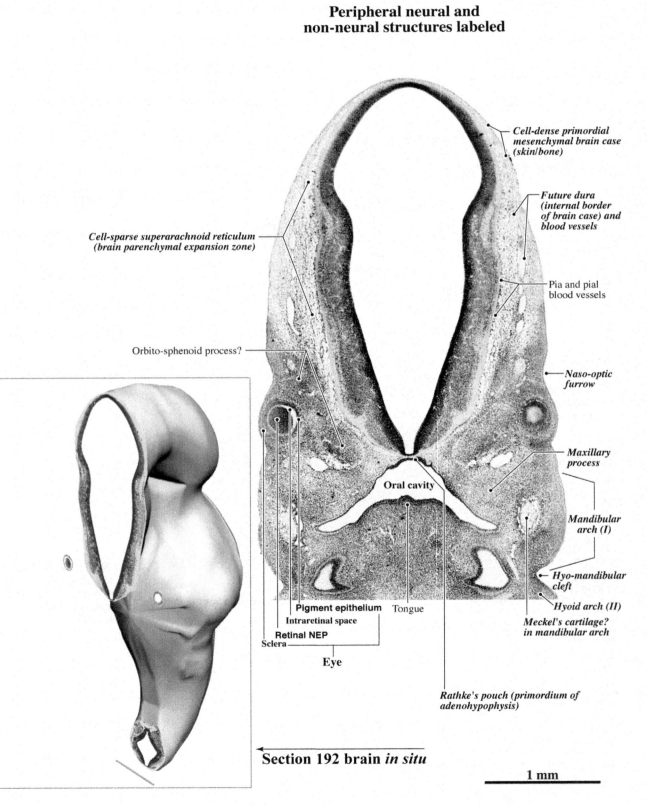

Cell-dense primordial mesenchymal brain case (skin/bone)

Future dura (internal border of brain case) and blood vessels

Pia and pial blood vessels

Cell-sparse superarachnoid reticulum (brain parenchymal expansion zone)

Naso-optic furrow

Orbito-sphenoid process?

Maxillary process

Oral cavity

Mandibular arch (I)

Hyo-mandibular cleft

Pigment epithelium　Tongue

Hyoid arch (II)

Intraretinal space

Retinal NEP

Meckel's cartilage? in mandibular arch

Sclera

Eye

Rathke's pouch (primordium of adenohypophysis)

← **Section 192 brain *in situ***

1 mm

Central neural structures labeled

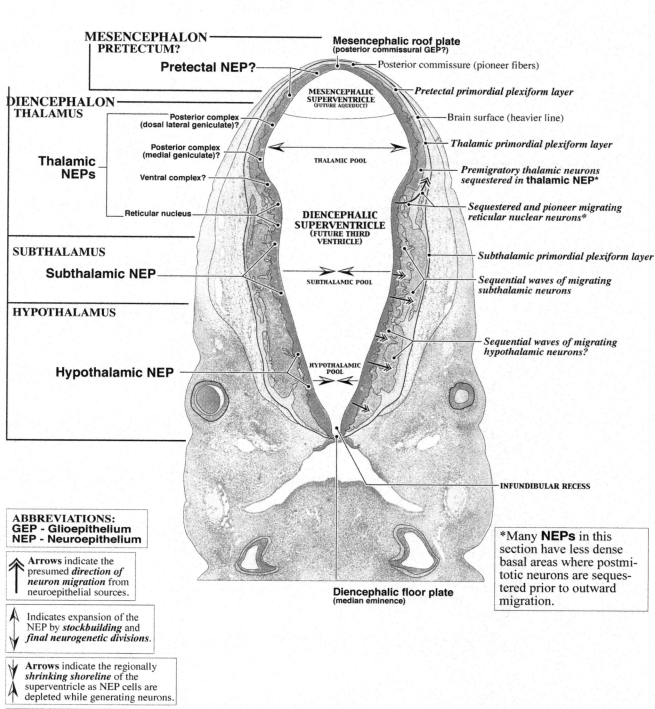

MESENCEPHALON ───────────── **Mesencephalic roof plate**
PRETECTUM? (posterior commissural GEP?)

Pretectal NEP?─── Posterior commissure (pioneer fibers)

MESENCEPHALIC SUPERVENTRICLE (FUTURE AQUEDUCT)

Pretectal primordial plexiform layer

DIENCEPHALON─────────
THALAMUS

Posterior complex (dosal lateral geniculate)?

Brain surface (heavier line)

Thalamic primordial plexiform layer

Thalamic NEPs

Posterior complex (medial geniculate)?

THALAMIC POOL

Premigratory thalamic neurons sequestered in **thalamic NEP***

Ventral complex?

Reticular nucleus

DIENCEPHALIC SUPERVENTRICLE (FUTURE THIRD VENTRICLE)

*Sequestered and pioneer migrating reticular nuclear neurons**

SUBTHALAMUS

Subthalamic NEP────

SUBTHALAMIC POOL

Subthalamic primordial plexiform layer

Sequential waves of migrating subthalamic neurons

HYPOTHALAMUS

Sequential waves of migrating hypothalamic neurons?

Hypothalamic NEP────

HYPOTHALAMIC POOL

INFUNDIBULAR RECESS

Diencephalic floor plate (median eminence)

ABBREVIATIONS:
GEP - Glioepithelium
NEP - Neuroepithelium

Arrows indicate the presumed *direction of neuron migration* from neuroepithelial sources.

Indicates expansion of the NEP by *stockbuilding* and *final neurogenetic divisions*.

Arrows indicate the regionally *shrinking shoreline* of the superventricle as NEP cells are depleted while generating neurons.

FONT KEY:
VENTRICULAR DIVISIONS – CAPITALS
Germinal zone - Helvetica bold
Transient structure - Times bold italic
Permanent structure - Times Roman or **Bold**

*Many **NEPs** in this section have less dense basal areas where postmitotic neurons are sequestered prior to outward migration.

PLATE 54A

CR 10 mm, GW6.5
M1000, Frontal/Horizontal
Section 215

Peripheral neural and
non-neural structures labeled

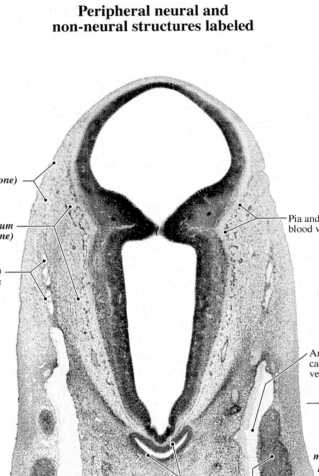

Cell-dense primordial mesenchymal brain case (skin/bone) —

Pia and pial
blood vessels

Cell-sparse superarachnoid reticulum
(brain parenchymal expansion zone)

Future dura (internal border of brain case)
and blood vessels

Anterior
cardinal
vein

Fused
maxillary
process
and
mandibular
arch (I)

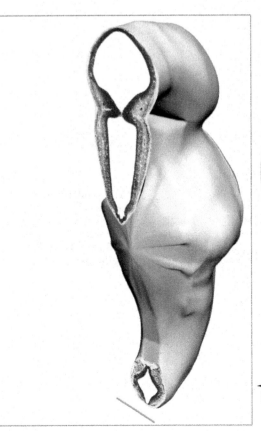

Trigeminal
ganglion (V)

Hyoid arch (II)

Rathke's pouch
(primordium of
adenohypophysis)

Nerve V (trigeminal)

Section 215 brain *in situ*

1 mm

Central neural structures labeled

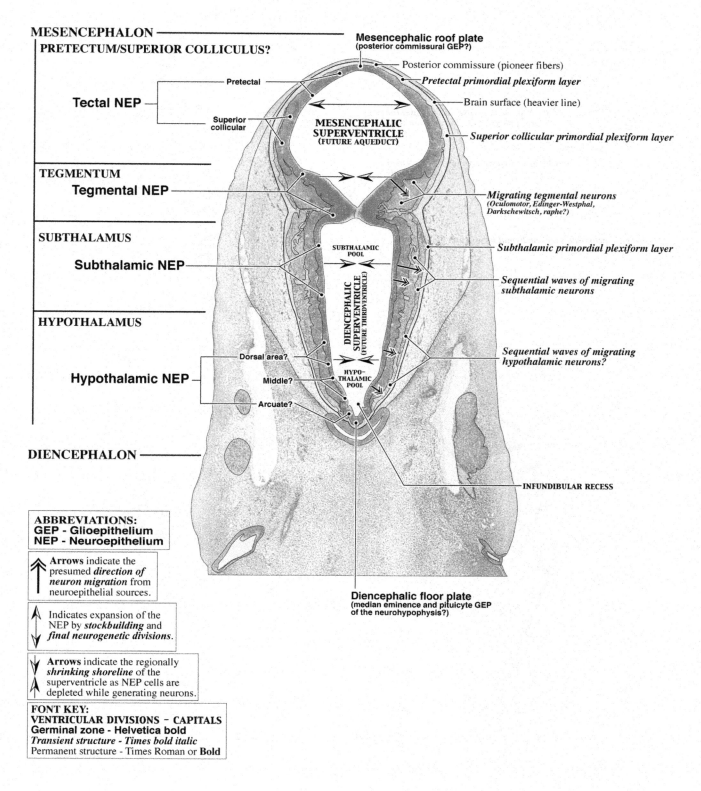

MESENCEPHALON
PRETECTUM/SUPERIOR COLLICULUS?

Tectal NEP

Pretectal

Superior
collicular

TEGMENTUM
Tegmental NEP

SUBTHALAMUS

Subthalamic NEP

HYPOTHALAMUS

Hypothalamic NEP

Dorsal area?

Middle?

Arcuate?

DIENCEPHALON

Mesencephalic roof plate
(posterior commissural GEP?)

Posterior commissure (pioneer fibers)

Pretectal primordial plexiform layer

Brain surface (heavier line)

MESENCEPHALIC
SUPERVENTRICLE
(FUTURE AQUEDUCT)

Superior collicular primordial plexiform layer

Migrating tegmental neurons
(Oculomotor, Edinger-Westphal,
Darkschewitsch, raphe?)

SUBTHALAMIC
POOL

DIENCEPHALIC
SUPERVENTRICLE
(FUTURE THIRDVENTRICLE)

Subthalamic primordial plexiform layer

Sequential waves of migrating
subthalamic neurons

HYPO-
THALAMIC
POOL

Sequential waves of migrating
hypothalamic neurons?

INFUNDIBULAR RECESS

Diencephalic floor plate
(median eminence and pituicyte GEP
of the neurohypophysis?)

ABBREVIATIONS:
GEP - Glioepithelium
NEP - Neuroepithelium

Arrows indicate the
presumed *direction of*
neuron migration from
neuroepithelial sources.

Indicates expansion of the
NEP by *stockbuilding* and
final neurogenetic divisions.

Arrows indicate the regionally
shrinking shoreline of the
superventricle as NEP cells are
depleted while generating neurons.

FONT KEY:
VENTRICULAR DIVISIONS - CAPITALS
Germinal zone - Helvetica bold
Transient structure - Times bold italic
Permanent structure - Times Roman or **Bold**

PLATE 55A

CR 10 mm, GW6.5
M1000, Frontal/Horizontal
Section 237

Peripheral neural and
non-neural structures labeled

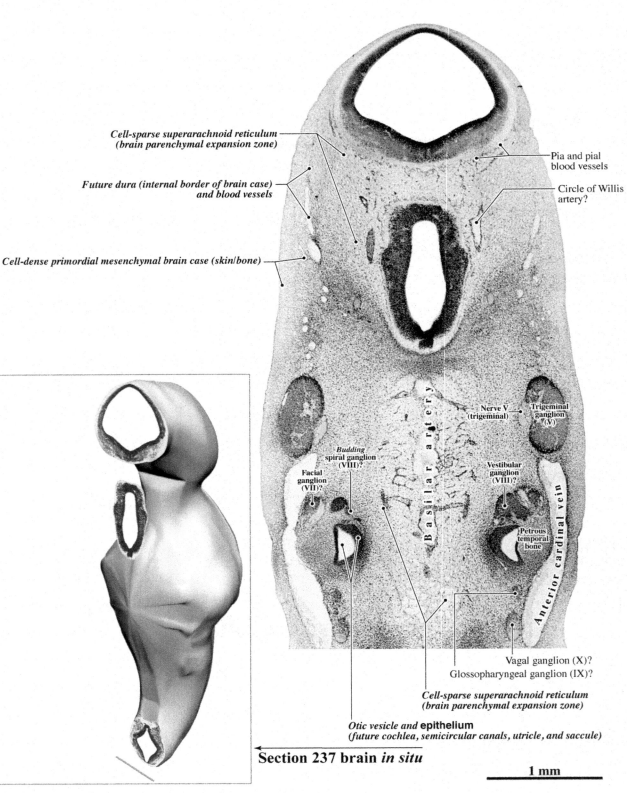

Cell-sparse superarachnoid reticulum
(brain parenchymal expansion zone)

Pia and pial
blood vessels

Future dura (internal border of brain case)
and blood vessels

Circle of Willis
artery?

Cell-dense primordial mesenchymal brain case (skin/bone)

Nerve V
(trigeminal)

Trigeminal
ganglion
(V)

Budding
spiral ganglion
(VIII)?

Vestibular
ganglion
(VIII)?

Facial
ganglion
(VII)?

B a s i l a r   a r t e r y

Petrous
temporal
bone

A n t e r i o r   c a r d i n a l   v e i n

Vagal ganglion (X)?
Glossopharyngeal ganglion (IX)?

Cell-sparse superarachnoid reticulum
(brain parenchymal expansion zone)

Otic vesicle and **epithelium**
(future cochlea, semicircular canals, utricle, and saccule)

Section 237 brain *in situ*

1 mm

Central neural structures labeled

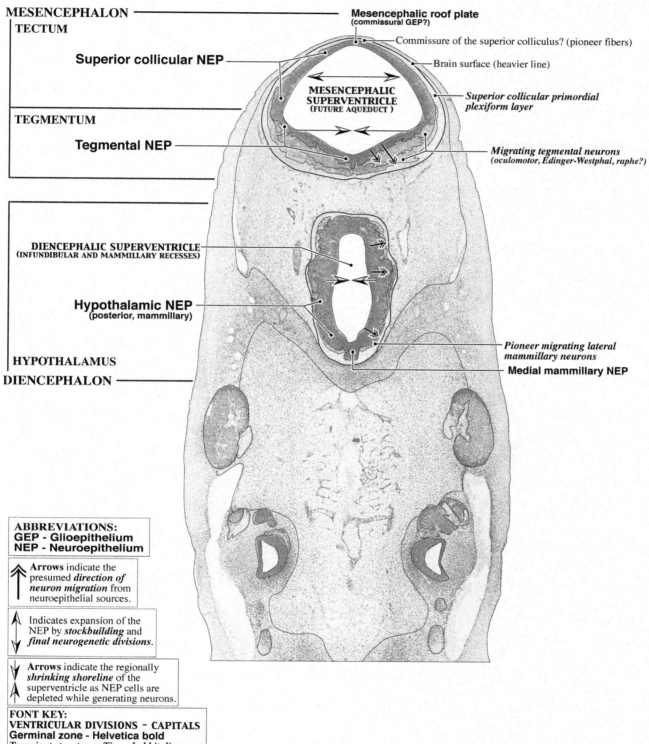

MESENCEPHALON ——————————————————— Mesencephalic roof plate
(commissural GEP?)
TECTUM
                                                        Commissure of the superior colliculus? (pioneer fibers)

Superior collicular NEP —————   Brain surface (heavier line)

                    ⟵————————⟶
                    MESENCEPHALIC
                    SUPERVENTRICLE   *Superior collicular primordial*
                    (FUTURE AQUEDUCT)   *plexiform layer*

TEGMENTUM

Tegmental NEP ———   *Migrating tegmental neurons*
                                                         (oculomotor, Edinger-Westphal, raphe?)

DIENCEPHALIC SUPERVENTRICLE
(INFUNDIBULAR AND MAMMILLARY RECESSES)

Hypothalamic NEP
(posterior, mammillary)

HYPOTHALAMUS   *Pioneer migrating lateral*
                                                         mammillary neurons
DIENCEPHALON ——————   Medial mammillary NEP

ABBREVIATIONS:
GEP - Glioepithelium
NEP - Neuroepithelium

Arrows indicate the
presumed *direction of*
neuron migration from
neuroepithelial sources.

Indicates expansion of the
NEP by *stockbuilding* and
final neurogenetic divisions.

Arrows indicate the regionally
shrinking shoreline of the
superventricle as NEP cells are
depleted while generating neurons.

FONT KEY:
VENTRICULAR DIVISIONS – CAPITALS
Germinal zone - Helvetica bold
Transient structure - Times bold italic
Permanent structure - Times Roman or **Bold**

PLATE 56A

CR 10 mm, GW6.5
M1000, Frontal/Horizontal
Section 255

**Peripheral neural
and non-neural
structures labeled**

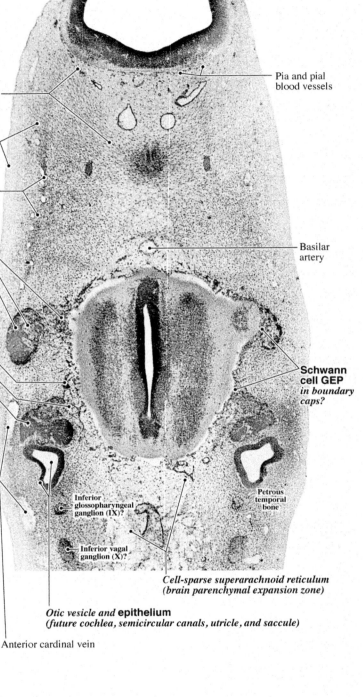

*Cell-sparse superarachnoid reticulum
(brain parenchymal expansion zone)*

Cell-dense primordial mesenchymal brain case (skin/bone)

*Future dura (internal border of brain case)
and blood vessels*

Nerve V *boundary cap*

Nerve V (trigeminal)

Trigeminal ganglion (V)

Nerve VIII *boundary cap*

Nerve VIII (vestibulocochlear)

Vestibulocochlear ganglion

Pia and pial
blood vessels

Basilar
artery

**Schwann
cell GEP**
*in boundary
caps?*

Inferior
glossopharyngeal
ganglion (IX)?

Inferior vagal
ganglion (X)?

Petrous
temporal
bone

*Cell-sparse superarachnoid reticulum
(brain parenchymal expansion zone)*

Otic vesicle and **epithelium**
(future cochlea, semicircular canals, utricle, and saccule)

Anterior cardinal vein

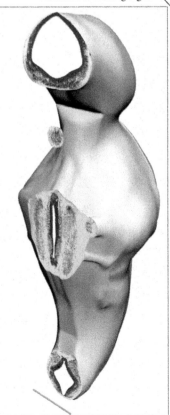

Section 255 brain *in situ*

1 mm

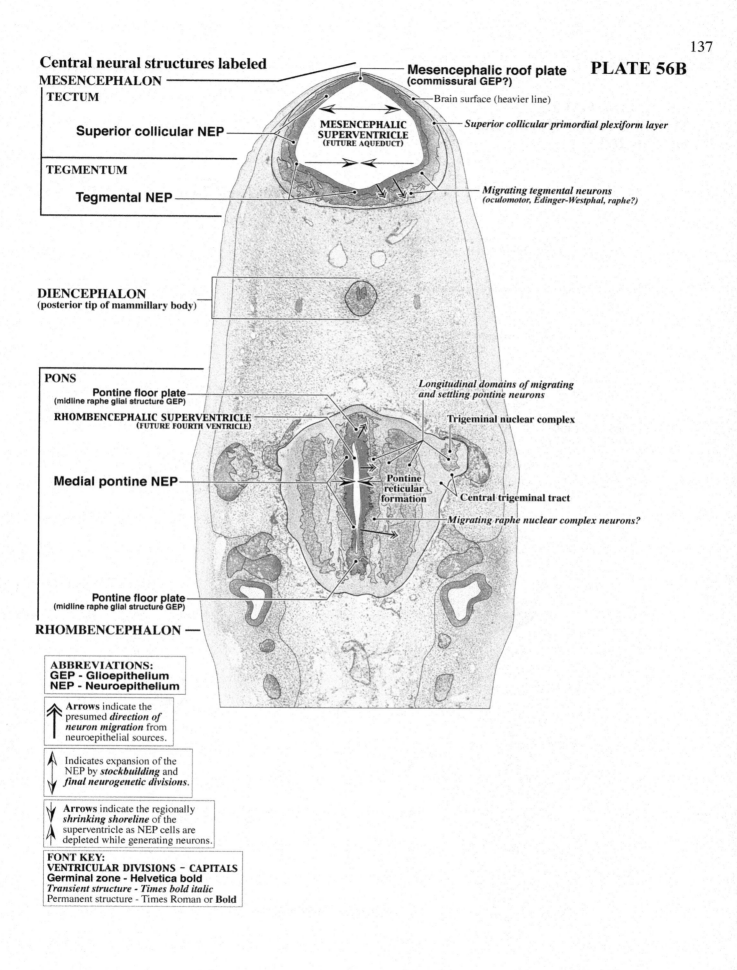

Central neural structures labeled

MESENCEPHALON ———————————————

Mesencephalic roof plate
(commissural GEP?)

PLATE 56B

TECTUM

Brain surface (heavier line)

Superior collicular NEP ———

MESENCEPHALIC
SUPERVENTRICLE
(FUTURE AQUEDUCT)

Superior collicular primordial plexiform layer

TEGMENTUM

Tegmental NEP ———

Migrating tegmental neurons
(oculomotor, Edinger-Westphal, raphe?)

DIENCEPHALON
(posterior tip of mammillary body)

PONS

Pontine floor plate
(midline raphe glial structure GEP)

Longitudinal domains of migrating
and settling pontine neurons

RHOMBENCEPHALIC SUPERVENTRICLE
(FUTURE FOURTH VENTRICLE)

Trigeminal nuclear complex

Medial pontine NEP ———

Pontine
reticular
formation

Central trigeminal tract

Migrating raphe nuclear complex neurons?

Pontine floor plate
(midline raphe glial structure GEP)

RHOMBENCEPHALON ———

ABBREVIATIONS:
GEP - Glioepithelium
NEP - Neuroepithelium

Arrows indicate the
presumed *direction of*
neuron migration from
neuroepithelial sources.

Indicates expansion of the
NEP by *stockbuilding* and
final neurogenetic divisions.

Arrows indicate the regionally
shrinking shoreline of the
superventricle as NEP cells are
depleted while generating neurons.

FONT KEY:
VENTRICULAR DIVISIONS – CAPITALS
Germinal zone - Helvetica bold
Transient structure - Times bold italic
Permanent structure - Times Roman or **Bold**

138

PLATE 57A

CR 10 mm, GW6.5
M1000, Frontal/Horizontal
Section 269

Peripheral neural and non-neural structures labeled

Pia and pial blood vessels

Cell-sparse superarachnoid reticulum (brain parenchymal expansion zone)

Cell-dense primordial mesenchymal brain case (skin/bone)

Future dura (internal border of brain case) and blood vessels

Basilar artery

Nerve V *boundary cap*

Schwann cell GEP *in boundary caps?*

Nerve VIII *boundary cap*

Nerve VIII (vestibulocochlear)

Vestibulocochlear ganglion (VIII)

Superior glossopharyngeal ganglion (IX)?

Superior vagal ganglion (X)?

Petrous temporal bone

Vertebral artery?

Cell-sparse superarachnoid reticulum (brain parenchymal expansion zone)

Otic vesicle and **epithelium** *(future cochlea, semicircular canals, utricle, and saccule)*

Section 269 brain *in situ*

1 mm

Central neural structures labeled

PLATE 57B

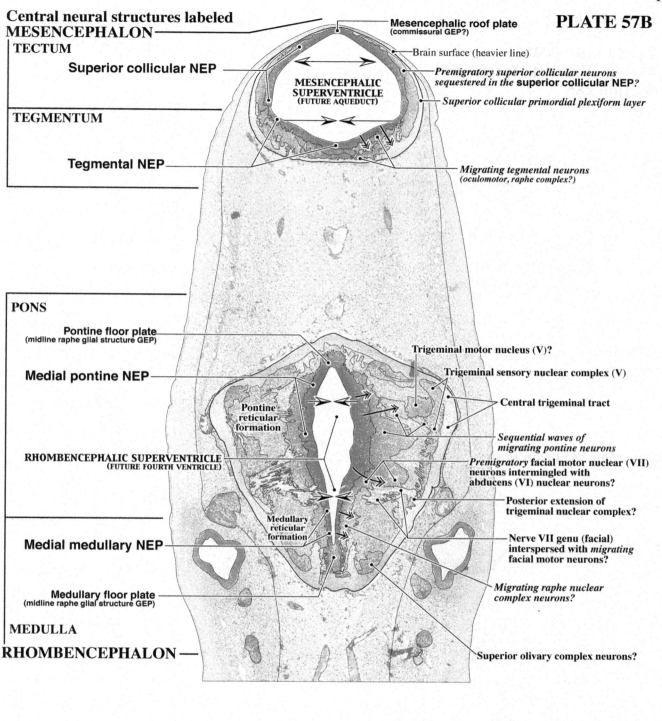

MESENCEPHALON ———

TECTUM

Superior collicular NEP

Mesencephalic roof plate
(commissural GEP?)

Brain surface (heavier line)

MESENCEPHALIC SUPERVENTRICLE
(FUTURE AQUEDUCT)

Premigratory superior collicular neurons
sequestered in the **superior collicular NEP***?*

Superior collicular primordial plexiform layer

TEGMENTUM

Tegmental NEP

Migrating tegmental neurons
(oculomotor, raphe complex?)

PONS

Pontine floor plate
(midline raphe glial structure GEP)

Medial pontine NEP

Pontine
reticular
formation

Trigeminal motor nucleus (V)?

Trigeminal sensory nuclear complex (V)

Central trigeminal tract

RHOMBENCEPHALIC SUPERVENTRICLE
(FUTURE FOURTH VENTRICLE)

Sequential waves of
migrating pontine neurons

Premigratory facial motor nuclear (VII)
neurons intermingled with
abducens (VI) nuclear neurons?

Posterior extension of
trigeminal nuclear complex?

Medial medullary NEP

Medullary
reticular
formation

Nerve VII genu (facial)
interspersed with *migrating*
facial motor neurons?

Medullary floor plate
(midline raphe glial structure GEP)

Migrating raphe nuclear
complex neurons?

MEDULLA

RHOMBENCEPHALON ——

Superior olivary complex neurons?

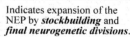

Arrows indicate the
presumed *direction of*
neuron migration from
neuroepithelial sources.

Indicates expansion of the
NEP by *stockbuilding* and
final neurogenetic divisions.

Arrows indicate the regionally
shrinking shoreline of the
superventricle as NEP cells are
depleted while generating neurons.

PLATE 58A

CR 10 mm, GW6.5
M1000, Frontal/Horizontal
Section 285

**Peripheral neural
and non-neural
structures labeled**

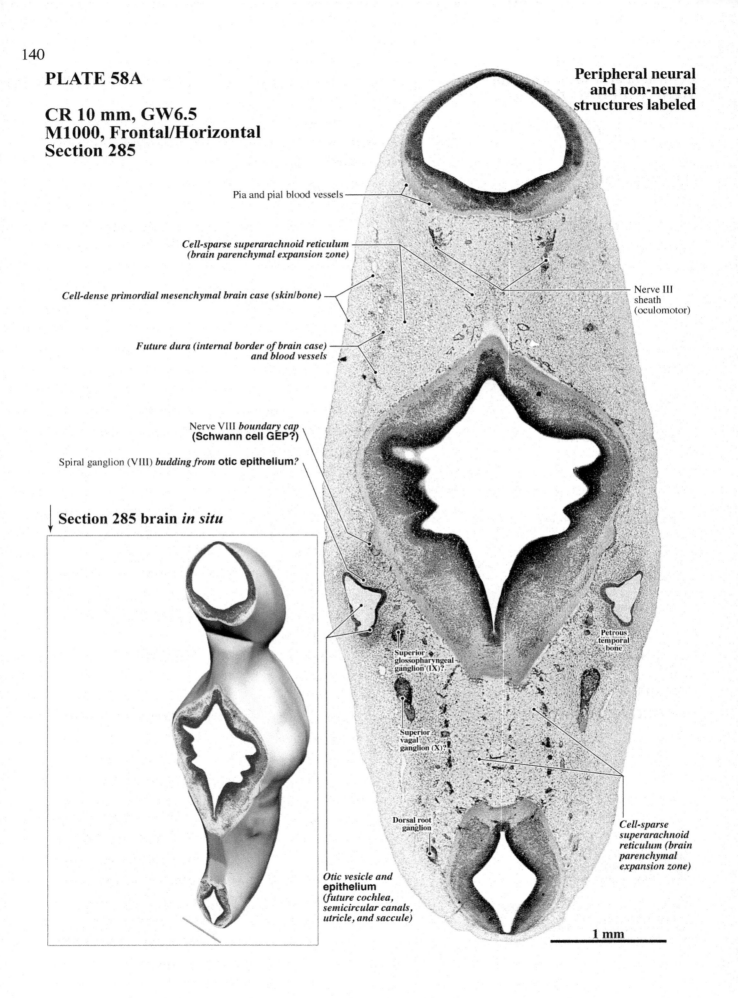

Pia and pial blood vessels

*Cell-sparse superarachnoid reticulum
(brain parenchymal expansion zone)*

Cell-dense primordial mesenchymal brain case (skin/bone)

*Future dura (internal border of brain case)
and blood vessels*

Nerve VIII *boundary cap*
(Schwann cell GEP?)

Spiral ganglion (VIII) *budding from* **otic epithelium?**

Section 285 brain *in situ*

Nerve III
sheath
(oculomotor)

Petrous
temporal
bone

Superior
glossopharyngeal
ganglion (IX)?

Superior
vagal
ganglion (X)?

Dorsal root
ganglion

*Cell-sparse
superarachnoid
reticulum (brain
parenchymal
expansion zone)*

Otic vesicle and
epithelium
*(future cochlea,
semicircular canals,
utricle, and saccule)*

1 mm

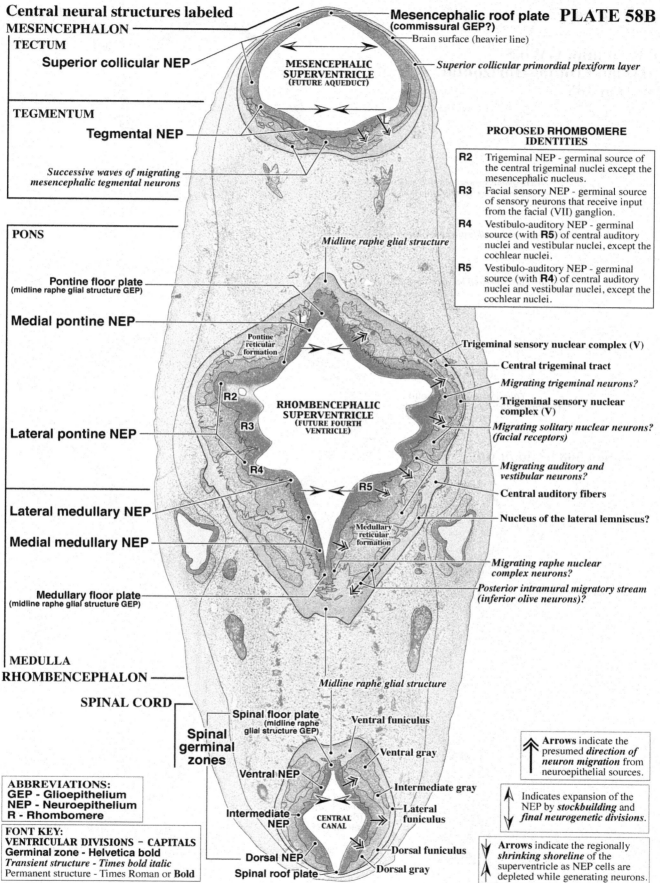

Central neural structures labeled **Mesencephalic roof plate** **PLATE 58B**
(commissural GEP?)

MESENCEPHALON
Brain surface (heavier line)

TECTUM
Superior collicular NEP
MESENCEPHALIC SUPERVENTRICLE (FUTURE AQUEDUCT)
Superior collicular primordial plexiform layer

TEGMENTUM
Tegmental NEP

Successive waves of migrating mesencephalic tegmental neurons

PROPOSED RHOMBOMERE IDENTITIES

R2 Trigeminal NEP - germinal source of the central trigeminal nuclei except the mesencephalic nucleus.
R3 Facial sensory NEP - germinal source of sensory neurons that receive input from the facial (VII) ganglion.
R4 Vestibulo-auditory NEP - germinal source (with **R5**) of central auditory nuclei and vestibular nuclei, except the cochlear nuclei.
R5 Vestibulo-auditory NEP - germinal source (with **R4**) of central auditory nuclei and vestibular nuclei, except the cochlear nuclei.

PONS

Midline raphe glial structure

Pontine floor plate
(midline raphe glial structure GEP)

Medial pontine NEP

Pontine reticular formation

Trigeminal sensory nuclear complex (V)
Central trigeminal tract
Migrating trigeminal neurons?
Trigeminal sensory nuclear complex (V)
Migrating solitary nuclear neurons? (facial receptors)

Lateral pontine NEP
RHOMBENCEPHALIC SUPERVENTRICLE (FUTURE FOURTH VENTRICLE)

R2 R3 R4 R5

Migrating auditory and vestibular neurons?
Central auditory fibers

Lateral medullary NEP
Medial medullary NEP
Medullary reticular formation
Nucleus of the lateral lemniscus?

Migrating raphe nuclear complex neurons?

Medullary floor plate
(midline raphe glial structure GEP)
Posterior intramural migratory stream (inferior olive neurons)?

MEDULLA
RHOMBENCEPHALON

Midline raphe glial structure

SPINAL CORD
Spinal floor plate
(midline raphe glial structure GEP)
Ventral funiculus
Spinal germinal zones
Ventral gray
Ventral NEP
Intermediate gray
Intermediate NEP
CENTRAL CANAL
Lateral funiculus
Dorsal funiculus
Dorsal NEP
Dorsal gray
Spinal roof plate

ABBREVIATIONS:
GEP - Glioepithelium
NEP - Neuroepithelium
R - Rhombomere

FONT KEY:
VENTRICULAR DIVISIONS – CAPITALS
Germinal zone - Helvetica bold
Transient structure - Times bold italic
Permanent structure - Times Roman or **Bold**

Arrows indicate the presumed *direction of neuron migration* from neuroepithelial sources.
Indicates expansion of the NEP by *stockbuilding* and *final neurogenetic divisions.*
Arrows indicate the regionally *shrinking shoreline* of the superventricle as NEP cells are depleted while generating neurons.

142

PLATE 59A

CR 10 mm, GW6.5
M1000, Frontal/Horizontal
Section 308

**Peripheral neural
and non-neural
structures labeled**

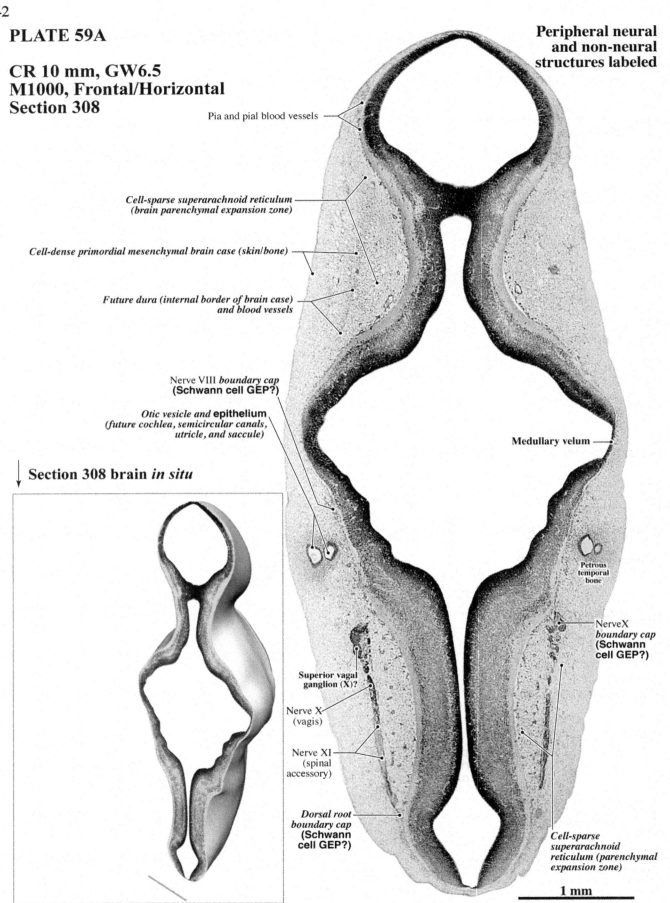

Pia and pial blood vessels

Cell-sparse superarachnoid reticulum
(brain parenchymal expansion zone)

Cell-dense primordial mesenchymal brain case (skin/bone)

Future dura (internal border of brain case)
and blood vessels

Nerve VIII *boundary cap*
(Schwann cell GEP?)

Otic vesicle and **epithelium**
(future cochlea, semicircular canals,
utricle, and saccule)

↓ Section 308 brain *in situ*

Medullary velum

**Petrous
temporal
bone**

Nerve X
boundary cap
**(Schwann
cell GEP?)**

Superior vagal
ganglion (X)?

Nerve X
(vagis)

Nerve XI
(spinal
accessory)

Dorsal root
boundary cap
(Schwann
cell GEP?)

Cell-sparse
superarachnoid
reticulum (parenchymal
expansion zone)

1 mm

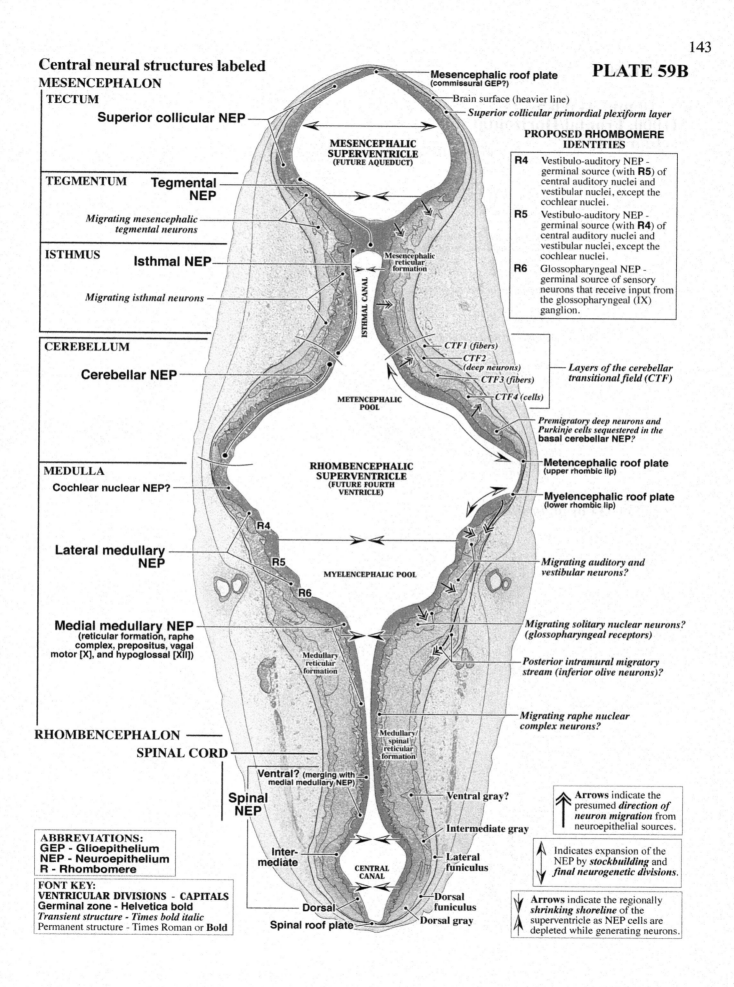

Central neural structures labeled

MESENCEPHALON

TECTUM

Superior collicular NEP

TEGMENTUM Tegmental NEP

Migrating mesencephalic tegmental neurons

ISTHMUS Isthmal NEP

Migrating isthmal neurons

CEREBELLUM

Cerebellar NEP

MEDULLA

Cochlear nuclear NEP?

Lateral medullary NEP

Medial medullary NEP
(reticular formation, raphe complex, prepositus, vagal motor [X], and hypoglossal [XII])

RHOMBENCEPHALON

SPINAL CORD

Spinal NEP

Mesencephalic roof plate
(commissural GEP?)

Brain surface (heavier line)

Superior collicular primordial plexiform layer

MESENCEPHALIC SUPERVENTRICLE (FUTURE AQUEDUCT)

Mesencephalic reticular formation

ISTHMAL CANAL

CTF1 (fibers)
CTF2 (deep neurons)
CTF3 (fibers)
CTF4 (cells)

Layers of the cerebellar transitional field (CTF)

METENCEPHALIC POOL

Premigratory deep neurons and Purkinje cells sequestered in the basal cerebellar NEP?

RHOMBENCEPHALIC SUPERVENTRICLE (FUTURE FOURTH VENTRICLE)

Metencephalic roof plate (upper rhombic lip)

Myelencephalic roof plate (lower rhombic lip)

R4 R5 R6

Migrating auditory and vestibular neurons?

MYELENCEPHALIC POOL

Migrating solitary nuclear neurons? (glossopharyngeal receptors)

Posterior intramural migratory stream (inferior olive neurons)?

Medullary reticular formation

Medullary/spinal reticular formation

Migrating raphe nuclear complex neurons?

Ventral? (merging with medial medullary NEP)

Ventral gray?

Intermediate gray

Intermediate

Lateral funiculus

CENTRAL CANAL

Dorsal

Dorsal funiculus
Dorsal gray

Spinal roof plate

PLATE 59B

PROPOSED RHOMBOMERE IDENTITIES

R4 Vestibulo-auditory NEP - germinal source (with **R5**) of central auditory nuclei and vestibular nuclei, except the cochlear nuclei.

R5 Vestibulo-auditory NEP - germinal source (with **R4**) of central auditory nuclei and vestibular nuclei, except the cochlear nuclei.

R6 Glossopharyngeal NEP - germinal source of sensory neurons that receive input from the glossopharyngeal (IX) ganglion.

ABBREVIATIONS:
GEP - Glioepithelium
NEP - Neuroepithelium
R - Rhombomere

FONT KEY:
VENTRICULAR DIVISIONS - CAPITALS
Germinal zone - Helvetica bold
Transient structure - Times bold italic
Permanent structure - Times Roman or **Bold**

Arrows indicate the presumed *direction of neuron migration* from neuroepithelial sources.

Indicates expansion of the NEP by *stockbuilding* and *final neurogenetic divisions*.

Arrows indicate the regionally *shrinking shoreline* of the superventricle as NEP cells are depleted while generating neurons.

144

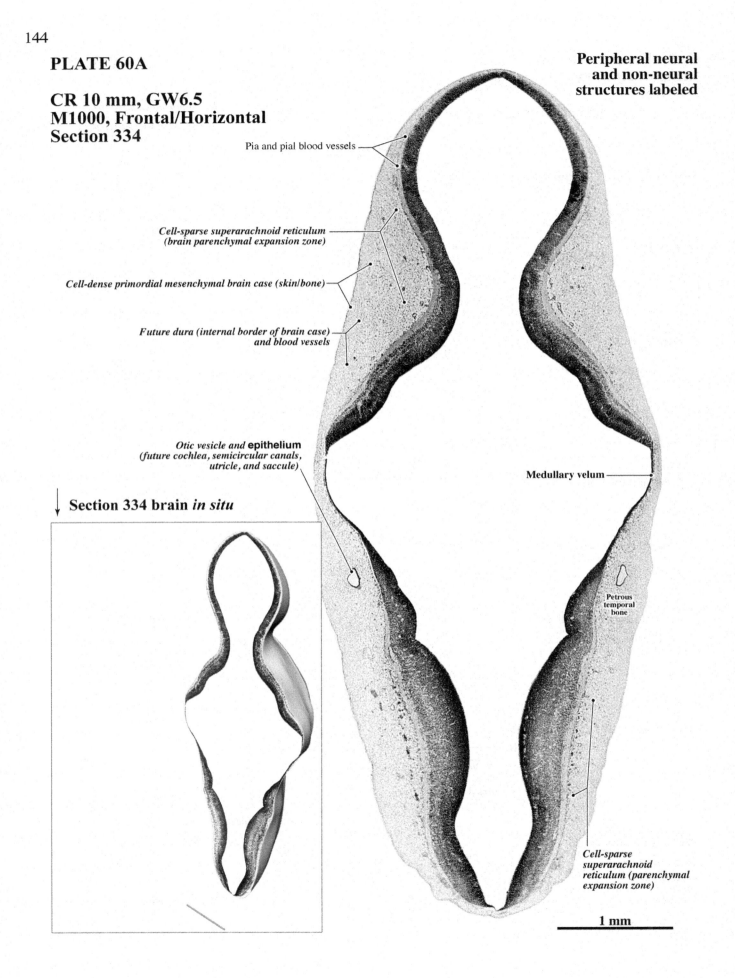

PLATE 60A

CR 10 mm, GW6.5
M1000, Frontal/Horizontal
Section 334

Peripheral neural
and non-neural
structures labeled

Pia and pial blood vessels

Cell-sparse superarachnoid reticulum
(brain parenchymal expansion zone)

Cell-dense primordial mesenchymal brain case (skin/bone)

Future dura (internal border of brain case)
and blood vessels

Otic vesicle and **epithelium**
(future cochlea, semicircular canals,
utricle, and saccule)

Medullary velum

Petrous
temporal
bone

↓ **Section 334 brain *in situ***

Cell-sparse
superarachnoid
reticulum (parenchymal
expansion zone)

1 mm

145

PLATE 60B

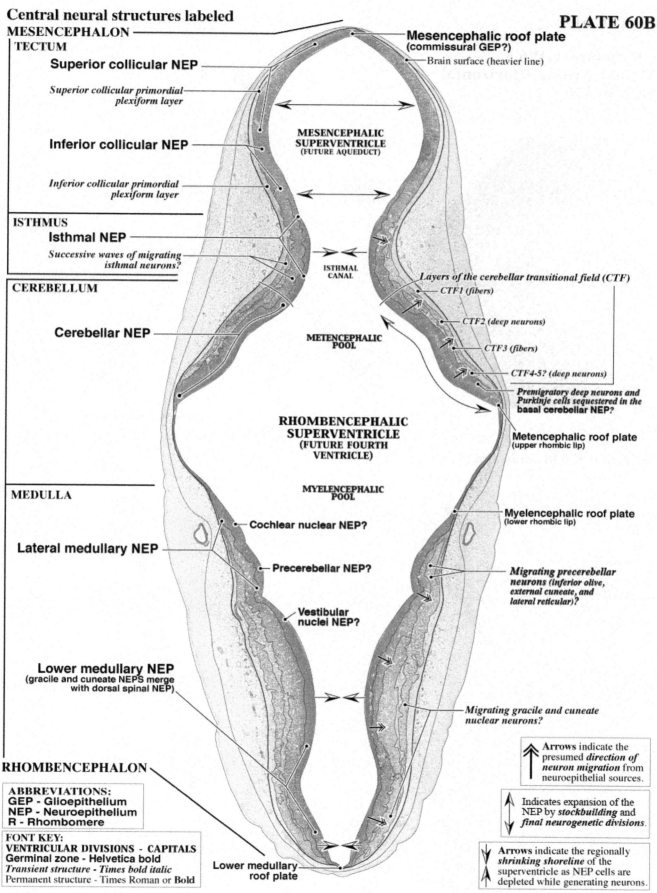

Central neural structures labeled

MESENCEPHALON
TECTUM
Superior collicular NEP
Superior collicular primordial plexiform layer
Inferior collicular NEP
Inferior collicular primordial plexiform layer
ISTHMUS
Isthmal NEP
Successive waves of migrating isthmal neurons?
CEREBELLUM
Cerebellar NEP
MEDULLA
Cochlear nuclear NEP?
Lateral medullary NEP
Precerebellar NEP?
Vestibular nuclei NEP?
Lower medullary NEP
(gracile and cuneate NEPS merge with dorsal spinal NEP)
RHOMBENCEPHALON

Mesencephalic roof plate
(commissural GEP?)
Brain surface (heavier line)
MESENCEPHALIC SUPERVENTRICLE (FUTURE AQUEDUCT)
ISTHMAL CANAL
Layers of the cerebellar transitional field (CTF)
CTF1 (fibers)
CTF2 (deep neurons)
METENCEPHALIC POOL
CTF3 (fibers)
CTF4-5? (deep neurons)
Premigratory deep neurons and Purkinje cells sequestered in the basal cerebellar NEP?
RHOMBENCEPHALIC SUPERVENTRICLE (FUTURE FOURTH VENTRICLE)
Metencephalic roof plate (upper rhombic lip)
MYELENCEPHALIC POOL
Myelencephalic roof plate (lower rhombic lip)
Migrating precerebellar neurons (inferior olive, external cuneate, and lateral reticular)?
Migrating gracile and cuneate nuclear neurons?
Lower medullary roof plate

ABBREVIATIONS:
GEP - Glioepithelium
NEP - Neuroepithelium
R - Rhombomere

FONT KEY:
VENTRICULAR DIVISIONS - CAPITALS
Germinal zone - Helvetica bold
Transient structure - Times bold italic
Permanent structure - Times Roman or **Bold**

Arrows indicate the presumed *direction of neuron migration* from neuroepithelial sources.

Indicates expansion of the NEP by *stockbuilding* and *final neurogenetic divisions*.

Arrows indicate the regionally *shrinking shoreline* of the superventricle as NEP cells are depleted while generating neurons.

PLATE 61A

**CR 10 mm, GW6.5
M1000, Frontal/Horizontal
Section 376**

**Peripheral neural
and non-neural
structures labeled**

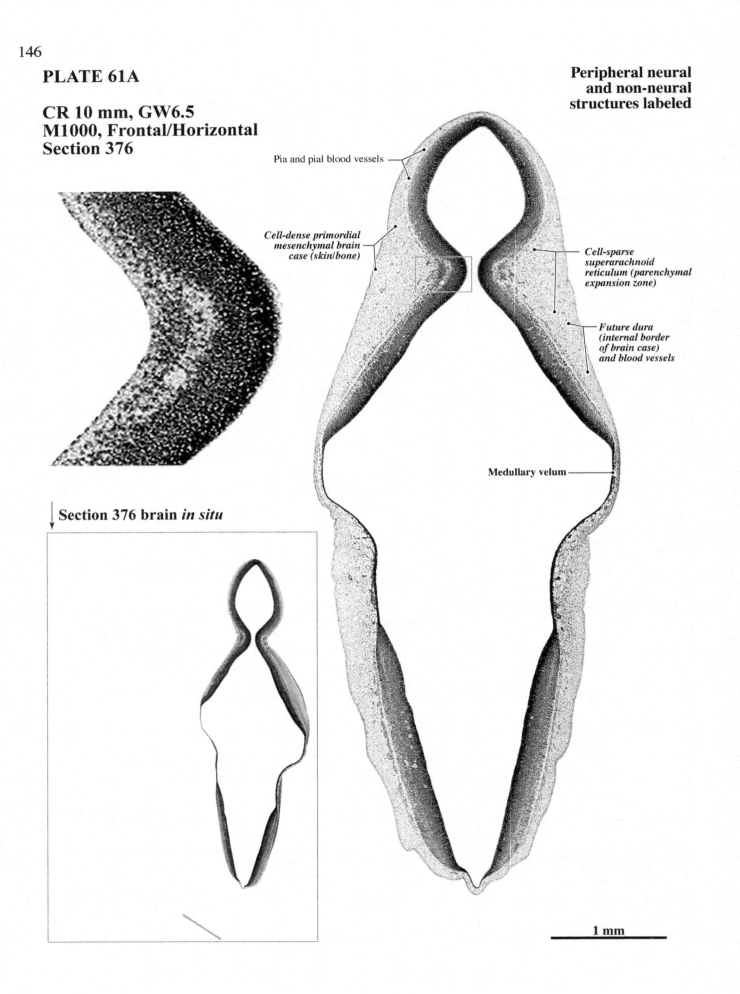

Pia and pial blood vessels

*Cell-dense primordial
mesenchymal brain
case (skin/bone)*

*Cell-sparse
superarachnoid
reticulum (parenchymal
expansion zone)*

*Future dura
(internal border
of brain case)
and blood vessels*

Medullary velum

Section 376 brain *in situ*

1 mm

Central neural structures labeled

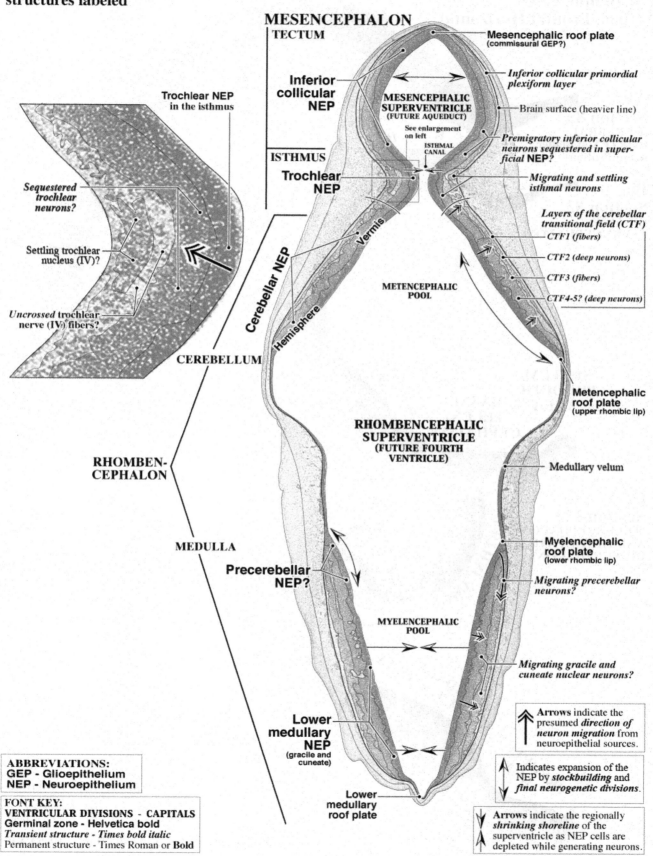

MESENCEPHALON

TECTUM

Inferior collicular NEP

ISTHMUS

Trochlear NEP

Vermis

Cerebellar NEP

Hemisphere

CEREBELLUM

Trochlear NEP in the isthmus

Sequestered trochlear neurons?

Settling trochlear nucleus (IV)?

Uncrossed trochlear nerve (IV₁) fibers?

RHOMBEN-CEPHALON

MEDULLA

Mesencephalic roof plate (commissural GEP?)

Inferior collicular primordial plexiform layer

MESENCEPHALIC SUPERVENTRICLE (FUTURE AQUEDUCT)

See enlargement on left

ISTHMAL CANAL

Brain surface (heavier line)

Premigratory inferior collicular neurons sequestered in superficial NEP?

Migrating and settling isthmal neurons

Layers of the cerebellar transitional field (CTF)

CTF1 (fibers)

CTF2 (deep neurons)

CTF3 (fibers)

CTF4-5? (deep neurons)

METENCEPHALIC POOL

RHOMBENCEPHALIC SUPERVENTRICLE (FUTURE FOURTH VENTRICLE)

Metencephalic roof plate (upper rhombic lip)

Medullary velum

Myelencephalic roof plate (lower rhombic lip)

Migrating precerebellar neurons?

Precerebellar NEP?

MYELENCEPHALIC POOL

Migrating gracile and cuneate nuclear neurons?

Lower medullary NEP (gracile and cuneate)

Lower medullary roof plate

Arrows indicate the presumed *direction of neuron migration* from neuroepithelial sources.

Indicates expansion of the NEP by *stockbuilding* and *final neurogenetic divisions*.

Arrows indicate the regionally *shrinking shoreline* of the superventricle as NEP cells are depleted while generating neurons.

ABBREVIATIONS:
GEP - Glioepithelium
NEP - Neuroepithelium

FONT KEY:
VENTRICULAR DIVISIONS - CAPITALS
Germinal zone - Helvetica bold
Transient structure - Times bold italic
Permanent structure - Times Roman or **Bold**

148

PLATE 62A
CR 10 mm, GW6.5
M1000, Frontal/Horizontal

A.
Section 85
(CEREBRAL
CORTEX at higher
magnification)

B.
Section 83
(ENTIRE
TELENCEPHALON)

CEREBRAL CORTEX

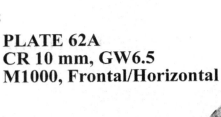

0.1 mm

SEPTUM/
PREOPTIC
AREA

BASAL
TELEN-
CEPHALON

BASAL
GANGLIA

0.25 mm

C.
Section 123
(TELENCEPHALON
AND PART OF THE
DIENCEPHALON)

SUBTHALAMUS

CEREBRAL CORTEX

AMYGDALA AND
BASAL GANGLIA

PREOPTIC
AREA

0.25 mm

FONT KEY:
VENTRICULAR DIVISIONS - CAPITALS
Germinal zone - Helvetica bold
Transient structure - Times bold italic
Permanent structure - Times Roman or **Bold**

See nearby entire sections in Plates 49-51AB.

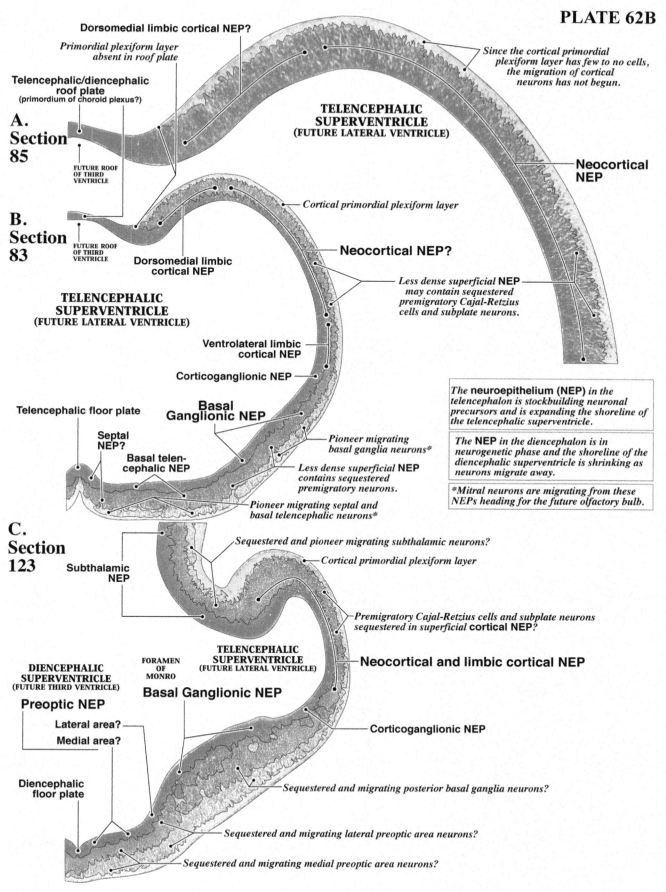

149

PLATE 62B

Dorsomedial limbic cortical NEP?

Primordial plexiform layer absent in roof plate

Telencephalic/diencephalic roof plate
(primordium of choroid plexus?)

A. Section 85

FUTURE ROOF OF THIRD VENTRICLE

Since the cortical primordial plexiform layer has few to no cells, the migration of cortical neurons has not begun.

TELENCEPHALIC SUPERVENTRICLE
(FUTURE LATERAL VENTRICLE)

Neocortical NEP

B. Section 83

FUTURE ROOF OF THIRD VENTRICLE

Dorsomedial limbic cortical NEP

Cortical primordial plexiform layer

Neocortical NEP?

Less dense superficial NEP may contain sequestered premigratory Cajal-Retzius cells and subplate neurons.

TELENCEPHALIC SUPERVENTRICLE
(FUTURE LATERAL VENTRICLE)

Ventrolateral limbic cortical NEP

Corticoganglionic NEP

Basal Ganglionic NEP

Telencephalic floor plate

Septal NEP?

Basal telencephalic NEP

*Pioneer migrating basal ganglia neurons**

Less dense superficial NEP contains sequestered premigratory neurons.

*Pioneer migrating septal and basal telencephalic neurons**

*The **neuroepithelium (NEP)** in the telencephalon is stockbuilding neuronal precursors and is expanding the shoreline of the telencephalic superventricle.*

*The **NEP** in the diencephalon is in neurogenetic phase and the shoreline of the diencephalic superventricle is shrinking as neurons migrate away.*

**Mitral neurons are migrating from these NEPs heading for the future olfactory bulb.*

C. Section 123

Sequestered and pioneer migrating subthalamic neurons?

Cortical primordial plexiform layer

Subthalamic NEP

Premigratory Cajal-Retzius cells and subplate neurons sequestered in superficial cortical NEP?

DIENCEPHALIC SUPERVENTRICLE
(FUTURE THIRD VENTRICLE)

FORAMEN OF MONRO

TELENCEPHALIC SUPERVENTRICLE
(FUTURE LATERAL VENTRICLE)

Neocortical and limbic cortical NEP

Preoptic NEP

Basal Ganglionic NEP

Lateral area?

Medial area?

Corticoganglionic NEP

Diencephalic floor plate

Sequestered and migrating posterior basal ganglia neurons?

Sequestered and migrating lateral preoptic area neurons?

Sequestered and migrating medial preoptic area neurons?